# A CONTINUING CHALLENGE IN COALITION AIR OPERATIONS

MYRON HURA
GARY McLEOD
ERIC LARSON
JAMES SCHNEIDER
DANIEL GONZALES
DANIEL NORTON
JODY JACOBS
KEVIN O'CONNELL
WILLIAM LITTLE
RICHARD MESIC
LEWIS JAMISON

RAND Project AIR FORCE

Prepared for the United States Air Force
Approved for public release; distribution unlimited

# PREFACE

This report describes research that was conducted (1) to help the U.S. Air Force identify potential interoperability problems that may arise in NATO Alliance operations or in U.S. coalition operations with NATO allies over the next decade; and (2) to suggest nonmateriel and technology-based solution directions to mitigate identified shortfalls. The focus of the research is on command, control, communications, intelligence, surveillance, and reconnaissance (C3ISR) systems and on out-of-NATO-area operations.

The research was sponsored by the Air Force Director of Intelligence, Surveillance, and Reconnaissance (USAF/XOI), the Air Force Director of Command and Control (USAF/XOC), and the commander of the Aerospace Command, Control, Intelligence, Surveillance, and Reconnaissance Center (AC2ISR/CC). The research was performed within the Aerospace Force Development program of Project AIR FORCE (PAF), and it builds on two recent PAF study projects: "Investment Guidelines for Information Operations—Focus on ISR" and "Developing Future Integrated C2 and ISR Capabilities."

This report should be of interest to policymakers, planners, and program managers involved in interoperability issues and programs of U.S. and NATO allies' air forces. It should also be of interest to planners and operational commanders involved in the employment of coalition C3ISR and combat capabilities.

## Project AIR FORCE

Project AIR FORCE, a division of RAND, is the Air Force federally funded research and development center (FFRDC) for studies and analysis. It provides the Air Force with independent analyses of policy alternatives affecting the development, employment, combat readiness, and support of current and future aerospace forces. Research is performed in four programs: Aerospace Force Development; Manpower, Personnel, and Training; Resource Management; and Strategy and Doctrine.

# CONTENTS

# FIGURES

# TABLES

# SUMMARY

The United States continues to invest in military capabilities to conduct unilateral operations if national interest so demands. At the same time, top-level national security and national military guidance and the preferences of top-level political and military decisionmakers increasingly require the U.S. military to participate in coalition operations. In some cases, coalition support is required for the United States to conduct successful military operations, and in most coalition operations the United States desires to share the burden. U.S. allies are also interested in coalitions because such operations provide them and with increased security and the opportunity to participate in military operations that the allies could not undertake unilaterally.

A key element in coalitions is interoperability. It enables allied support for coalition operations and can increase the effectiveness and efficiency of U.S. and allied forces in such operations. However, because a predominantly technical treatment of interoperability cannot cover certain strategic and operational implications, the research described in this report uses a broad definition that is common to the U.S. Department of Defense and to NATO:

> *Interoperability.* The ability of systems, units, or forces to provide services to and accept services from other systems, units, or forces, and to use the services so exchanged to enable them to operate effectively together.[1]

---

[1]See *Joint Staff* (1999).

## RESEARCH OBJECTIVE

The objective of this research is to help the U.S. Air Force identify potential interoperability problems that may arise in coalition air operations of the United States and NATO allies over the next decade, and to suggest nonmateriel and technology-based solution directions to address identified challenges. The focus of the research is on command, control, communications, intelligence, surveillance, and reconnaissance (C3ISR) systems in out-of-NATO-area operations. For purposes of this study, we use a broader definition of "out-of-NATO-area" (or "out-of-area") operations. Our definition includes NATO Alliance operations and non-NATO coalition operations in which the United States and other NATO allies participate and that occur outside or on the periphery of Alliance members' territory. Although the authors recognize that interoperability problems remain in joint-service operations and are worthy of research, such problems are beyond the scope of the effort documented here.

## STUDY APPROACH

To better understand interoperability and its multiple dimensions, we conducted literature reviews, had discussions with subject matter experts, and surveyed 40 recent coalition operations (one of which was a NATO Alliance out-of-area operation). To identify interoperability challenges that the United States and its NATO allies' air forces will need to address over the next decade, we examined new technology, security, and military trends as reflected in top-level DoD guidance and visions of future military operations.

Next we identified and conducted case studies of key ongoing U.S. and allied programs that have major implications for future interoperability. To better understand the military value of these programs, we examined them in the context of conducting representative military missions. In parallel, we examined the economic and political implications of interoperability for the selected programs to determine those approaches that offer the best opportunity to address interoperability issues.

Based on these findings, we developed suggested solution directions, both nonmateriel and technology-based, to address identified interoperability challenges.

## OBSERVATIONS

### Recent Operations

Our review of recent coalition operations indicates that interoperability has multiple and complex dimensions—political and economic as well as military—that may manifest themselves at strategic, operational, tactical, and technological levels. Further, interoperability problems are not isolated by level. Strategic-level interoperability problems can have operational and tactical implications, and technological interoperability problems may reverberate in the opposite direction. For example, the political and economic goals of individual nations to support national industries can lead to the development of air power systems (e.g., fighters, weapons, airborne surveillance and control assets) that have substantially different capabilities and that require extensive work-arounds to be employed in coalition operations. Similarly, the lack of interoperable communications and combat identification systems and procedures could result in the attrition of coalition aircraft to enemy defenses or fratricide that causes the partner to leave the coalition. Interoperability issues should be considered in the context of each of these levels.

Political support, access to allied infrastructures and airspace, landing rights, and forward basing are essential to bringing U.S. air power to bear effectively. Specifically, allied support for and actual participation in coalitions help U.S. decisionmakers garner and maintain the public support necessary to conduct military operations in regions of the world that are of national interest (e.g., Southwest Asia [SWA] and the Balkans). Moreover, as seen, for example, in Operation Allied Force, access to allied air space and the availability of infrastructure in close proximity to areas of operations can minimize flight time to air patrol stations and targets while providing flexibility to conduct attack operations from more than one approach azimuth.

These factors may be overlooked if the potential contributions of individual allies are measured solely in terms of the number of aircraft made available or sorties flown in specific operations. Although it is true that in recent SWA operations (e.g., Desert Storm, Desert Fox) the United States provided the preponderance of air missions, in some Balkan coalition operations—particularly those not involving precision strike operations (e.g., air surveillance, air space control, and no-fly-zone enforcement)—allies provided more than half of the aircraft and missions flown. Further, it is important to recognize that providing the preponderance of air assets to coalition operations helped the United States rationalize its smaller ground force contributions in Bosnia and Kosovo operations.

Notwithstanding the above, allied contributions to recent strike operations in the Balkans have been limited because the allies lack sufficient precision-guided weapons (PGWs) that can be delivered day or night in any weather conditions. A number of U.S. allies are developing plans to expand their holdings of PGWs, including those that use the Global Positioning System (GPS). These plans need to be implemented because future crises may require the use of PGWs to minimize casualties and collateral damage as well as standoff weapons to minimize the risk of attrition of coalition aircraft to more sophisticated enemy air defenses.

Another future concern lies in the allies' limited capabilities (e.g., force readiness, airlift) to rapidly deploy forces to out-of-area operations. The allies have made great strides in their ability to deploy and support operations outside their borders to the periphery of Europe, but more improvement is needed to rapidly deploy combat forces to non-European theaters. Without improvements to existing capabilities, the combat value of allied air forces is likely to decrease.

Recognizing that these and other shortfalls exist among its members, NATO endorsed the Defence Capabilities Initiative in April 1999 to meet the challenges of the present and foreseeable security environment. The most important areas NATO identified for improvement were the deployability and mobility of Alliance forces, their sustainability and logistics, their survivability and effective engagement capability, and the necessary command and control and information systems. These improvements are needed not only for future NATO air operations but also for other likely coalition air operations.

### Case Studies

The case studies examined suggest that the following areas offer the best leverage for achieving acceptable levels of interoperability in future coalition operations: (1) a common or harmonized doctrine for the planning, execution, execution monitoring, and assessment of combined joint task force operations, especially air campaigns; (2) compatible or adaptable concepts of operations and procedures for airborne surveillance and control in support of air-to-air and air-to-ground missions; (3) common information-sharing standards and compatible tactical communication systems; and (4) expert personnel who understand the capabilities of coalition partners and who hone their expertise in combined operations and exercises.

Efforts to enhance interoperability solely through common or fully interoperable systems at the technological level are likely to be limited by political, economic, and security factors, particularly the desire to support national industries, equitable burden sharing, and ensuring that the most advanced military capabilities are not compromised. From a technology and cost perspective, selected C3ISR initiatives appear to offer the best opportunities for interoperability enhancements.

## SUGGESTED ACTIONS

Our suggested actions fall into two categories: those that the U.S. Air Force can take only in collaboration with other stakeholders and those for which the Air Force has purview to take direct action. Whereas the United States is developing capabilities to conduct unilateral operations if national interest demands, most NATO allies are developing their capabilities to conduct operations in the context of NATO concepts, processes, and systems. Thus, the suggested actions also consider relevant NATO-wide developments.

### Collaborative Actions

In collaboration with the DoD, other U.S. services, and NATO allies, the U.S. Air Force should

- Help NATO develop the Combined Joint Task Force (CJTF) concept of operations (CONOPS), associated processes, expert personnel, systems, and information-sharing protocols for out-of-area operations. In particular, the Air Force should ensure that the key doctrinal concept of centralized control and decentralized execution, which is inherent in U.S. joint-service air CONOPS, is institutionalized in the NATO CJTF concept.
- Help NATO define the desired level of information sharing and interoperability between planned U.S. and NATO force-level planning and execution-monitoring capabilities (organizations, procedures, personnel, and systems).[2] At a minimum, a set of common messaging standards for information exchange should be defined for the U.S. Air Force's Theater Battle Management Core System (TBMCS) and NATO's Interim Combined Air Operations Center (CAOC) Capability as well as for TBMCS and NATO's Air Command and Control System.
- Help NATO develop a coherent space policy and information-sharing protocols that provide sufficient information to conduct key operations while protecting sensitive equities. In some cases, bilateral agreements with selected NATO allies may be appropriate.
- Continue to foster the interoperability of Airborne Warning and Control System (AWACS) assets and standard air early warning and control procedures, especially those needed in the presence of friendly and enemy stealth aircraft. The focus should be on ensuring that NATO, U.K., French, and U.S. AWACS modernization programs are synchronized.
- Develop the process and capabilities to receive and exploit ground surveillance information from the different airborne ground moving-target indication (GMTI) sensors that NATO members are developing. Support advanced concept demonstrations to determine the value of this capability and help select

---

[2]We focus on interoperability between U.S. and NATO capabilities because only a few NATO countries have extensive national air command and control capabilities: not only do most NATO allies rely on NATO-wide capabilities, they rationalize their spending by investing in such capabilities.

the most appropriate means to achieve it, including development of common GMTI data formats.[3]

- Ensure that the Multifunctional Information Distribution System (MIDS) engineering and manufacturing development program is successfully completed and that the functional interoperability inherent in MIDS terminals is maintained through the production phase and then applied to future fighter data links.
- Continue to share fighter and weapon systems information to ensure adequate common understanding of individual coalition partners' air capabilities (technology, personnel, operations). In parallel, continue to develop operating protocols that permit the use of allied air assets in coalition operations and expand training exercises to emphasize out-of-area operations. Be prepared to employ workarounds.
- Encourage NATO allies' acquisition of advanced precision weapons and standoff weapons. Low-cost GPS-guided weapons are particularly promising. Although they are expensive, standoff weapons ensure platform survivability in a high-threat environment, and standoff antiarmor weapons enable more effective participation in the halt phase of a campaign.
- Increase opportunities for combined experiments and advanced technology demonstrations.
- Support the above suggested actions by actively participating in NATO's Defence Capabilities Initiative.

## Direct Actions

In parallel with the preceding collaborative efforts, the U.S. Air Force should consider taking the following direct actions:

- Leverage its expertise and capabilities in planning and executing air and space operations in power projection missions by manning key positions in the emerging deployable and key static CAOCs to reinforce the principle of centralized control and

---

[3]These are appropriate actions given the uncertainties of the NATO Alliance Ground Surveillance program.

decentralized execution. Further, it should develop and maintain a cadre of experts who can provide support to higher NATO headquarters (if needed) to help develop air campaign plans and assist in execution monitoring and assessment.

- Explore opportunities to gain better visibility into the Western European Union Satellite Centre to determine if and how Centre assets might help satisfy some of the information needs in future NATO operations.
- Ensure that the AWACS Radar System Improvement Program (RSIP) continues to be adequately funded and that appropriate NATO RSIP employment lessons learned are incorporated in future early warning and air control doctrine and tactics.
- Support advanced concept technology demonstration of multiple GMTI sensor data reception and exploitation capabilities in joint expeditionary force experiments.
- Strengthen Air Force visibility and management oversight in the MIDS production-phase program to ensure that MIDS terminals are delivered as needed to U.S. fighter modernization programs, within budget constraints.

## IN SHORT

The United States and the U.S. Air Force can influence its NATO allies up to a point, recognizing that self-interest will remain paramount. At the strategic level, the allies do not put high-intensity conflict at the centerpiece of their planning. They do not see a superpower threat to NATO arising or any serious military threat to their well-being. Hence, their strategic focus is on peace operations and crisis response. The result is proportionately lower investment relative to the United States in developing and acquiring advanced military systems such as stealth aircraft and all-weather PGWs.

This suggests that efforts to develop nonsystem solutions are likely to be more successful in improving the interoperability of U.S. and allied air forces. Offering the most promise are the development of agreed-upon information-sharing protocols and means; force and tactical employment concepts, processes, and procedures (e.g., development, dissemination, and execution of air tasking order); and expert personnel capable of instituting workarounds for interoperability problems that will continue to arise in operational and tactical levels of coalition operations.

Because of their much lower relative cost and greater flexibility in application, technological solutions to interoperability that are based on C3ISR elements appear more feasible than attempts to achieve interoperability through the acquisition of common major weapon systems (such as high-performance aircraft). Among the most promising are interoperable airborne surveillance and control capabilities and tactical data communications systems to support air-to-air and air-to-ground operations.

# ACKNOWLEDGMENTS

We gratefully acknowledge the steadfast support we received from our study points of contact, Captain Dean Adkins (AF/XOII) and Major Jackson Harris (AF/XOCI) at the Air Staff and Lieutenant Colonel George Caragianis (C2RS) at the Aerospace C2 & ISR Center (AC2ISRC). We also thank Colonel Ray Briscoe (C2R), Colonel Jack Fellows (C2G), and Colonel Joseph May (C2C) and their staff at the AC2ISRC for providing useful and relevant information as well as for arranging visits to Air Force and NATO organizations.

To conduct the interoperability case studies, RAND researchers visited a number of organizations, including Department of Defense organizations, NATO organizations, foreign defense and military organizations, and U.S. defense contractors. We would like to thank the following individuals either for arranging the visits and/or for providing information used in the case studies:

SHAPE Headquarters: Lieutenant Colonel Al Schaake (USAF) and Wing Commander Hilary Whiteway (U.K. Royal Air Force).

NATO Command, Control, and Consultation Agency: Lieutenant Colonel Paul Dundas (USAF), Lieutenant Colonel Nancy Deming (USAF), Joseph Rodero, Joe Ross, and Lucius Strazdis (MITRE).

NATO Air Command and Control Management Agency: Thomas Corbitt.

NATO Airborne Early Warning and Control Programme Management Agency: Thomas Brownell, George Wolfe, and Fletcher Thomson.

NATO Airborne Early Warning Force Command: Colonel Clark Wigley (USAF), Lieutenant Colonel David Wininger (USAF), Gordon Parkhill.

NATO Interim Deployable Combined Air Operations Center (CAOC, Ramstein): Lieutenant Colonel Gottlieb Ohl (German Air Force).

Office of the Secretary of Defense: Dr. Judith Daly (ODUSD[AS&C]) and Robert Bruce (ODUSD(I&CP)).

Joint Strike Fighter Program Office: Lieutenant Colonel William Shelton (USAF) and Colonel Jouke Eikelboom (Denmark Air Force).

Multifunctional Information Distribution System International Program Office: Captain Tom Russell (USN) and Colonel Craig Christen (USAF).

Joint Staff/J-6: Lieutenant Colonel Charles Murray (USAF).

Office of the Secretary of the Air Force: Major Jim Ashworth (AQI).

USAF Electronic Systems Command: Lieutenant Colonel Greg Juday (AWACS SPO), Major Kirk Streitmater (JSTARS SPO), "Hamp" Huckins (JSTARS SPO), George Lewis (Tactical Data Link SIO), and Randy Burnham (Tactical Data Link SIO).

U.K. Defence Evaluation and Research Agency: Dr. David Hutber (Malvern) and Dr. David Owen (Centre for Defence Analysis).

German Air Force Flying Training Center: Major Axel Meierhoefer.

French Ministry of Defense: Colonel Emmanuel de Romemont (French Air Force).

French Air Force: Colonel Joel Martel.

U.S. Defense Contractors: Mr. Bruce Long (Lockheed-Martin), Pat Murphy (Lockheed-Martin), Tom O'Lear (Lockheed-Martin), Pamela Roose (Motorola), and Robert Kirch (Motorola).

RAND researchers also visited United States Air Forces in Europe (USAFE) Headquarters, Ramstein Air Base, Germany, and the CAOC at Vicenza, Italy, to discuss interoperability issues that arose during

Operation Allied Force. We spoke to many U.S. and NATO personnel too numerous to mention by name. However, we would like to thank Major Glenn Best (USAFE/DOQT) for arranging our visit to USAFE and Major Mark Buccigrossi (USAF) for arranging our visit to the CAOC.

Several RAND colleagues made valuable contributions to this research. Lieutenant Colonel Jim Keefer (USAF) and Lieutenant Colonel Ken Gardiner (USMC), as FY 1999 Military Fellows at RAND, and Gustav Lindstrom researched a number of recent coalition operations and identified key interoperability challenges. John Baker provided useful insights regarding current space developments. Recent work on the interdiction of moving armor columns by David Ochmanek, Edward Harshberger, and Carl Rhodes provided useful data and context for our halt analysis. Finally, we would like to thank Jimmie McEver for use of his computer model for the halt analysis. Of course, we alone are responsible for any errors of omission or commission.

# ACRONYMS

| | |
|---|---|
| AAA | Antiaircraft artillery |
| AB | Air base |
| ABCCC | Airborne Battlefield Command and Control Center |
| ABL | Airborne laser |
| ACCS | Air Command and Control System |
| ACO | Airspace control order |
| ACTD | Advanced concept technology demonstration |
| AEW | Airborne early warning |
| AEW&C | Airborne early warning and control |
| AFB | Air Force Base |
| AFV | Armored fighting vehicle |
| AGS | Alliance Ground Surveillance |
| AIP | ASARS 2 Improvement Program |
| AOC | Air operations center |
| AODB | Air Operations Database |
| ARL | Airborne Reconnaissance Low |
| ASEAN | Association of Southeast Asian Nations |
| ASOC | Air support operations center |
| ASTOR | Airborne Stand-Off Radar |
| ATACMS | Army Tactical Missile System |
| ATO | Air tasking order |
| AWACS | Airborne Warning and Control System |
| BAI | Battlefield air interdiction |

| | |
|---|---|
| BDA | Bomb damage assessment |
| BICES | Battlefield Information Collection and Exploitation Systems |
| BM | Battle management |
| BVRAAM | Beyond Visual Range Air-to-Air Missile |
| C2 | Command and control |
| C3 | Command, Control, and Communications |
| C3ISR | Command, control, communications, intelligence, surveillance, and reconnaissance |
| C4ISR | Command, control, communications, computers, intelligence, surveillance, and reconnaissance |
| CAESAR | Coalition Aerial Surveillance and Reconnaissance |
| CAOC | Combined air operations center |
| CAP | Combat air patrol |
| CC | Component Commander |
| CC&D | Camouflage, concealment, and detection |
| CEB | Combined effects bomblet |
| CEC | Cooperative Engagement Capability |
| CINC | Commander-in-chief |
| CIS | Combat Information System |
| CIWS | Close-In Weapon System |
| CJTF | Combined Joint Task Force |
| CM | Cruise Missile |
| CNAD | Conference of NATO Armaments Directors |
| CNES | Centre National d'Études Spatiales [French Space Agency] |
| COE | Common operating environment |
| COLE | Concept of Link 16 Employment |
| ComCJTF | CJTF commander |
| CONOPS | Concept of operations |
| CONUS | Continental United States |
| COP | Common operational picture |

| | |
|---|---|
| COSMO | Constellation of [Small] Satellites for Mediterranean [Basin] Observation |
| CRC | Control and reporting center |
| CRE | Control and reporting element |
| CRESO | Complesso Radar Eliportato per la Sorveglianza [Italian helicopter-based GMTI system] |
| CRONOS | Crisis Response Operation in NATO Open System |
| CTAPS | Contingency Theater Air Planning System |
| CTP | Common tactical picture |
| DAPE | Dynamic assessment, planning, and execution |
| DCA | Defensive counterair |
| DCAOC | Deployable CAOC |
| DDR&E | Director Defense Research and Engineering |
| DEM | Digital elevation model |
| DGA | Délégation Général pour l'Armament |
| DGS | Deployable ground station |
| DII | Defense Information Infrastructure |
| DLI | Data link infrastructure |
| DLWG | Data Link Working Group |
| DoD | Department of Defense |
| DTDMA | Distributed time division multiple access |
| EAPC | Euro-Atlantic Partnership Council |
| ECM | Electronic countermeasures |
| EGNOS | European Geostationary Overlay Service |
| EMD | Engineering and manufacturing development |
| ERS | European Remote Sensing |
| ESA | European Space Agency |
| ESC | Electronic Systems Center |
| ESDI | European Security and Defence Identity |
| ESM | Electronic support measures |
| EU | European Union |
| FDL | Fighter Data Link |

| | |
|---|---|
| FYROM | Former Yugoslav Republic of Macedonia |
| GCCS | Global Command and Control System |
| GEO | Geostationary orbit |
| GINS | GPS Integrated Navigation System |
| GLONASS | Russian navigation satellite system |
| GMTI | Ground moving target indication |
| GNSS 2 | Global Navigation Satellite System 2 (aka Galileo navigation satellite system) |
| GPS | Global Positioning System |
| GRCA | Ground radar coverage area |
| HARM | High-Speed Anti-Radiation Missile |
| HORIZON | Helicoptère d'Observation Radar et d'Investigation de Zone [French helicopter-based GMTI system] |
| HQ | Headquarters |
| HRR | High range resolution |
| HTML | Hypertext markup language |
| HTS | HARM Targeting System |
| I&W | Indications and warning |
| IBS | Integrated Broadcast Service |
| ICC | Interim CAOC Capability |
| IEEE | Institute of Electrical and Electronic Engineers |
| IFF | Identification friend or foe |
| IFOR | Implementation Force |
| IFV | Infantry fighting vehicle |
| IJMS | Interim Joint Message Standard |
| INS | Inertial navigation system |
| I/O | Input/output |
| IOC | Initial operational capability |
| IORG | Interoperability Review Group |
| IP | Industrial participation |
| IPO | International Program Office |
| IRWG | International Requirements Working Group |

| | |
|---|---|
| ISR | Intelligence, surveillance, and reconnaissance |
| JADO | Joint air defense operations |
| JASSM | Joint Air-to-Surface Standoff Missile |
| JCTN | Joint composite tracking network |
| JDAM | Joint Direct Attack Munition |
| JDN | Joint data network |
| JEFX | Joint expeditionary force experiment |
| JEZ | Joint engagement zone |
| JIRCRB | JTIDS International Configuration Review Board |
| JMOA | Joint memorandum of agreement |
| JMTOP | Joint Multi-TADIL Operating Procedures |
| JOA | Joint operational area |
| JSF | Joint Strike Fighter |
| JSOW | Joint Standoff Weapon |
| JSRC | Joint Sub-Regional Command |
| JSTARS | Joint Surveillance [and] Target Attack Radar System |
| JTIDS | Joint Tactical Information Distribution System |
| JTRS | Joint Tactical Radio System |
| JV 2010 | *Joint Vision 2010* |
| LANTIRN | Low-Altitude Navigation and Targeting Infrared System for Night |
| LINUX | A version of the UNIX computer operating system |
| LOC | Line of communication |
| LOCAAS | Low-Cost Anti-Armor System |
| LOCE | Linked Operational Intelligence Centers Europe |
| LRIP | Low-rate initial production |
| LVT | Low Volume Terminal |
| MACC | Multinational AEW Commanders Conference |
| MCCIS | Maritime Command and Control Information System |
| MEO | Medium earth orbit |
| MIDB | Modernized Integrated Database |

| | |
|---|---|
| MIDS | Multifunctional Information Distribution System |
| MISREP | Mission report |
| MLU | Mid-Life Upgrade |
| MMP | Mid-Term Modernisation Programme |
| MOOTW | Military operations other than war |
| MOR | Military operational requirement |
| MPCD | Multipurpose color display |
| MSI | Multi-Sensor Integration |
| MSS | Mobile satellite system |
| MTW | Major theater war |
| NAC | North Atlantic Council |
| NAEWF | NATO Airborne Early Warning Force |
| NAEWFC | NATO Airborne Early Warning Force Command |
| NAFAG | NATO Air Force Armaments Group |
| NAPMA | NATO AEW&C Programme Management Agency |
| NAPMO | NATO AEW&C Programme Management Organisation |
| NATO | North Atlantic Treaty Organization |
| NC3A | NATO Consultation, Command, and Control Agency |
| NIST | National Institute of Standards and Technology |
| NMP | Near-Term Modernisation Programme |
| NTE | Not to exceed |
| OASD(C3I) | Office of the Assistant Secretary of Defense (Command, Control, Communications, and Intelligence) |
| OCA | Offensive counterair |
| OSD | Office of the Secretary of Defense |
| OT&E | Operational test and evaluation |
| PAA | Primary aircraft authorized |
| PAC-3 | Patriot Advanced Capability-3 |
| PEC | Program Executive Council |

| | |
|---|---|
| PEO | Program Executive Officer |
| PfP | Partnership for Peace |
| PGM | Precision-guided munition |
| PGW | Precision-guided weapon |
| PIE | Paris Interoperability Experiment |
| PMOU | Program memorandum of understanding |
| PNT | Positioning, navigation, and timing |
| RC | Regional Command |
| RCS | Radar cross section |
| RDT&E | Research, development, test, and evaluation |
| ROE | Rules of engagement |
| RSIP | Radar System Improvement Program |
| RTIP | Radar Technology Insertion Program |
| SACEUR | Supreme Allied Commander Europe |
| SACLANT | Supreme Allied Commander Atlantic |
| SAM | Surface-to-air missile |
| SAR | Synthetic aperture radar |
| SATCOM | Satellite communications |
| SDC | Situation display console |
| SEAD | Suppression of enemy air defenses |
| SEM-E | Standard Electronic Modules Format-E |
| SFOR | Stabilization Force |
| SFW | Sensor-Fuzed Weapon |
| SHAPE | Supreme Headquarters Allied Powers Europe |
| SHAR | [Rockwell-Collins] Sea Harrier [terminal] |
| SI AHWG | SOSTAS Interoperability Ad Hoc Working Group |
| SIAP | Single integrated air picture |
| SIAR WG | Standing Interoperability and Applications Working Group on Intelligence, Surveillance, and Reconnaissance |
| SIO | Systems Integration Office [at Electronic Systems Center] |

| | |
|---|---|
| SOSTAR | Standoff Surveillance and Target Acquisition Radar |
| SOSTAS | Standoff Surveillance and Target Acquisition Systems |
| SPO | System program office |
| SPOT | Satellite pour l'Observation de la Terre [French Earth observation satellite] |
| SSS | System Segment Specification |
| STANAG | Standardization agreement |
| SWA | Southwest Asia |
| TACS | Theater Air Control System |
| TADIL | Tactical digital information link |
| TAMD | Theater air and missile defense |
| TBM | Theater ballistic missile |
| TBMCS | Theater Battle Management Core System |
| TDMA | Time division multiple access |
| TDP | Technical Data Package |
| TEL | transporter-erector-launcher |
| TLE | target location error |
| TPFDD | Time-phased force deployment document |
| TTP | Tactics, techniques, and procedures |
| UNSCOM | United Nations Special Commission |
| USAFE | United States Air Forces in Europe |
| WCMD | Wind-Corrected Munitions Dispenser |
| WEU | Western European Union |
| WMD | Weapons of mass destruction |

Chapter One

# INTRODUCTION

## STUDY OBJECTIVE AND SCOPE

The objective of this research is twofold: (1) to help the U.S. Air Force identify potential interoperability problems that may arise in coalition air operations of the United States and other willing NATO allies over the next decade,[1] and (2) to suggest solution directions to mitigate those problems. The research develops both nonmateriel and technology solutions to address identified shortfalls, taking into consideration force structure elements, doctrine and tactics, and interoperability workarounds.

The focus of the research is on command, control, communications, intelligence, surveillance, and reconnaissance (C3ISR) systems in out-of-NATO-area operations. For purposes of this study, we use a broader definition of "out-of-NATO-area" (or "out-of-area") operations. Our definition includes NATO Alliance operations and non-NATO coalition operations in which the United States and other NATO allies participate and that occur outside or on the periphery of Alliance territory.

---

[1]Although the authors recognize that interoperability problems remain in joint-service operations and are worthy of research, they are beyond the scope of the effort documented here.

## BACKGROUND

The United States continues to invest in military capabilities to conduct unilateral operations. At the same time, top-level national security and national military guidance and the preferences of top-level political and military decisionmakers increasingly require the U.S. military to participate in coalition operations.

Interoperability is a key element in coalitions because it enables allied support for coalition operations and offers the opportunity to increase the effectiveness and efficiency of U.S. and allied forces in such operations. Achieving these potential benefits, however, is not a trivial matter. Interoperability problems among coalition air forces may arise for a variety of reasons beyond the usual technical problems encountered, for example, in radio communications. Differences may exist in the speed at which coalition air forces can plan operations or can deploy to areas where there are mutual interests to protect. For example, the U.S. Air Force is reorganizing itself to be more expeditionary and to be able to deliver bombs on target within 48 hours of receiving the order to deploy from the continental United States (CONUS). Also, there may be differences in doctrine or employment concepts. For example, U.S. allies may not place the same emphasis that the United States places on speedy deployment, precision strike, or the halt phase of a major theater war (MTW). These doctrinal and operational differences can be compounded by technical differences among forces.

The planned modernization of the U.S. Air Force's fighter and bomber force will render that force both far more stealthy and capable of precise delivery of weapons from standoff ranges against a spectrum of targets. In parallel, the U.S. Air Force, in collaborative efforts with other services, the Department of Defense (DoD), and other government organizations, is upgrading its C3ISR capabilities to support a modernized fighter and bomber force. U.S. allies are not likely to follow suit to the same extent. This divergence in capabilities between the United States and its allies is becoming more apparent and must be properly managed to ensure that the potential benefits of coalition operations are realized.

## STUDY APPROACH

We began the study by discussing the many definitions of interoperability and underscoring its inherent multiple dimensions and complex relationships. Based on literature reviews, discussions with subject matter experts, and several of the authors' direct encounters with interoperability issues in coalition operations, we highlighted its importance in the evolving context of new security, budgetary, and programmatic issues.

We continued with a review of recent coalition operations to better understand the dimensions, issues, and value of interoperability. Analyses and documented lessons learned from 40 recent operations involving U.S. and NATO allies' air forces, together with discussions with subject matter experts, form the basis for this step of our research. The results of these efforts, and our examination of new technology, security, and military trends as reflected in top-level DoD guidance and visions of future military operations, were used to identify interoperability challenges that U.S. and NATO allies' air forces will need to address over the next decade.

Next, we identified and conducted case studies of key ongoing U.S. and allied programs with important implications for future interoperability. The case studies encompass five major elements necessary to conduct air operations: (1) concepts (doctrine, processes, systems, and personnel) for force level command and control (C2), assessment, planning, and execution; (2) information-sharing practices and systems; (3) airborne surveillance and control for air-to-air and air-to-ground operations; (4) tactical digital data communications; and (5) weapon systems. The case studies included an examination of economic and political implications to determine approaches that offer the best opportunity to address interoperability issues.

To cover this broad range of elements within the resources of this study, we focused on selected programs of direct interest to the Air Force. Note that our examination addressed interoperability issues and not whether the specific program or programs examined provide the optimal solution for conducting particular functions of air operations.

To better understand their military value, we examined these selected programs in the context of conducting representative military missions. Mission-level and campaign-level analyses were used to assess the value of interoperability.

Based on these findings, we developed suggested solution directions to address identified interoperability challenges. We subdivided the solution directions into two categories: (1) those that need to be done in collaboration with other major stakeholders, and (2) those for which the U.S. Air Force can choose to take direct action.

## STRUCTURE OF THE REPORT

In Chapter Two, we discuss definitions of interoperability and underscore its inherent multiple dimensions and complex relationships. In Chapter Three, we summarize our review of recent coalition operations. In Chapter Four, we identify interoperability challenges that U.S. and NATO allies' air forces will need to address over the next decade. These three chapters also provide the context for the case studies discussed in the next six chapters.

Chapters Five through Ten provide case studies and suggested actions that the U.S. Air Force can take to realize short- and medium-term improvements in the interoperability of U.S. and NATO allies' air forces. The case study in Chapter Five examines air command and control in the context of the recent NATO reorganization and its expansion from Article 5 defense of NATO territory to a more power projection–oriented C2 structure to support out-of-area operations. We focus on interoperability between U.S. and NATO capabilities because only a few NATO countries have extensive national air command and control capabilities.

The case study on space development (Chapter Six) highlights potential cooperation or competition between U.S. space programs that were vital in past coalitions and Europe's growing desire to increase its space capabilities and hence lessen its dependence on U.S. assets in future military operations.

Chapters Seven and Eight look at case studies of airborne surveillance and control platforms. Air surveillance and control capability—based on U.S., NATO, U.K., and French Airborne

Warning and Control System (AWACS) programs—is viewed as a major interoperability success. By contrast, ground surveillance and control—based on the Joint Surveillance [and] Target Radar System (JSTARS), NATO's Alliance Ground Surveillance concept, and U.K., French, and Italian programs—remains an interoperability challenge.

Chapter Nine examines the use of tactical data links, in particular Link 16, for achieving interoperability among U.S. and NATO allies' fighters. The case study focuses on a near-term fighter data link program, the Multifunctional Information Distribution System (MIDS), a cooperative program between five NATO nations and the United States to develop a low-cost Link 16 terminal. The MIDS case study highlights the programmatic complexities of international cooperative initiatives.

Chapter Ten examines the interoperability among U.S. and NATO allies' fighters and weapon systems. With the development of multinational European fighter aircraft, future coalition operations will be characterized by less commonality between U.S. and NATO allies' fighter forces, which could create interoperability challenges for force planners. Limited development and procurement of precision-guided weapons (PGWs) by U.S. NATO allies create additional challenges.

In Chapter Eleven, we illustrate the military value of interoperability. We analyze air surveillance during peacekeeping operations, force protection against conventional aircraft and cruise missiles using defensive counterair (DCA) capabilities, and interdiction of moving columns of armor during the halt phase of an MTW. For each of these missions, we describe the operational concept, identify the system capabilities of the NATO ally participants, and highlight actual and potential contributions of allied forces in coalition operations with the United States.

In Chapter Twelve, we summarize our observations and present suggested actions to improve interoperability of U.S. and NATO allies' air forces.

Appendix A provides a brief summary of our analyses of 40 recent coalition operations involving U.S. and NATO allies' air forces. This is background material for the discussions presented in Chapters

Three through Ten. A separate report presents in more detail our analysis of recent operations.[2]

Appendix B briefly describes the four new operational concepts presented in *Joint Vision 2010* as background material for Chapter Four.[3]

Appendix C provides details of the MIDS case study presented in Chapter Nine.

Appendix D summarizes our analysis of the deployment of additional U.S. Air Force fighter forces to theater to augment in-place fighter forces. In the scenario analyzed, the forces are being used to interdict moving armor columns as part of a halt-the-invasion operation. We consider two cases: in one the U.S. deployment is fully supported by NATO allies; in the other there is no support from NATO allies. The results are used as inputs to the halt analysis presented in Chapter Eleven.

---

[2]See Larson et al. (1999).

[3]See Chairman of the Joint Chiefs of Staff (1996).

Chapter Two

# A BROAD DEFINITION OF INTEROPERABILITY

*Interoperability* would seem to be a straightforward concept. Put simply, is a measure of the degree to which various organizations or individuals are able to operate together to achieve a common goal. From this top-level perspective, interoperability is a good thing, with overtones of standardization, integration, cooperation, and even synergy.

Interoperability specifics, however, are not well defined. They are often situation-dependent, come in various forms and degrees, and can occur at various levels—strategic, operational, and tactical as well as technological. They are also far more likely to be recognized when interoperability problems emerge and taken for granted when such problems do not.

Interoperability often comes at a price. These costs may be difficult to define and estimate insofar as they consist of military expenditures to enhance interoperability as well as the economic and political costs incurred. The issue, of course, is what sorts of interoperability are worth what sorts of costs.

Because of these various levels and multiple dimensions, we examine interoperability from the broadest available definition:

> The ability of systems, units, or forces to provide services to and accept services from other systems, units, or forces, and to use the

services so exchanged to enable them to operate effectively together.[1]

This broad definition of interoperability encompasses several areas that narrower definitions may not, including (1) the ability of forces from different nations to work effectively together given the nature of the forces and the combined military organizational structure (the traditional narrow sense); (2) the effectiveness of the combined military organizational structure (e.g., how well can the C2 structure allocate combined assets to achieve military goals); and (3) the degree of similarity of technical capabilities of the forces from different nations, reflecting their fungibility in supporting coalition military goals (e.g., do they have similar precision strike capabilities?).

Thus, this broad definition lets us look at interoperability in all its dimensions and offers the promise of revealing the most feasible and prudent interoperability-enhancing options: those options that address the most pressing problems while minimizing the costs to NATO allies and to the United States.

In the remainder of this chapter, we elaborate on this definition by examining interoperability in greater detail in the context of four levels—strategic, operational, tactical, and technological (as depicted in Figure 2.1 in the context of conducting an air campaign).

## STRATEGIC PERSPECTIVES

At the strategic level, interoperability is an enabler for coalition building. It facilitates meaningful contributions by coalition partners. It supports whatever allied "buy-in" may be necessary for the United States to use its forces effectively in regions of interest. As the current formulation of the United States' national security strategy states:

---

[1]Host nation support such as communication networks, infrastructure, air bases, and aircraft squadrons and special forces are examples of services, units, and forces. See Joint Staff (1999), p. 229.

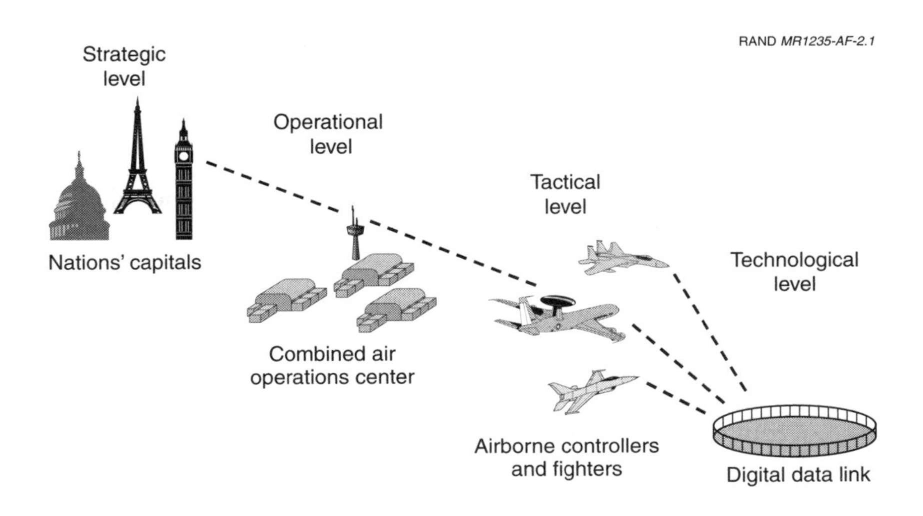


**Figure 2.1—Interoperability Examined at Four Levels**

> We must always be prepared to act alone when that is our most advantageous course. But many of our security objectives are best achieved—or can only be achieved—through our alliances and other formal security structures, or as a leader of an ad hoc coalition formed around a specific objective. Durable relationships with allies and friendly nations are vital to our security. A central thrust of our strategy is to strengthen and adapt the security relationships we have with key nations around the world and create new relationships and structures when necessary. Examples include NATO enlargement, the Partnership for Peace, the NATO-Russia Permanent Joint Council, the African Crisis Response Initiative, the regional security dialogue in the ASEAN [Association of Southeast Asian Nations] Regional Forum and the hemispheric security initiatives adopted at the Summit of the Americas.[2]

At the highest level, interoperability issues center on harmonizing the world views, strategies, doctrines, and force structures of the United States and its allies (in this study, NATO members). Interoperability at this level is an element of alliance/coalition will-

---

[2]See The White House (1998), p. 2.

ingness to work together over the long term to achieve and maintain shared interests (e.g., adherence to the rule of international law, democracy, human rights, and open markets) against common threats. As such, interoperability provides a measure of deterrence to would-be troublemakers, and it helps motivate and shape defense research and development, acquisition, strategy, doctrine, tactics, training, and combined exercises.

Alliance and coalition interoperability is one means of achieving both effective and efficient military capabilities: a rationalized approach to interoperability can reduce alliance-wide military expenditures, increase the flexibility or fungible character of selected forces, or define military niches that will be provided by national members so that redundancy can be avoided. Further, substantial participation in coalitions can increase burden sharing by spreading both the costs and risks that are incurred.

The top-level DoD vision for future warfighting concepts, *Joint Vision 2010* (JV 2010), describes the importance of interoperability in multinational operations as follows:

> It is not enough to be joint, when conducting future operations. We must find the most effective methods for integrating and improving interoperability with allied and coalition partners. Although our Armed Forces will maintain decisive unilateral strength, we expect to work in concert with allied and coalition forces in nearly all of our future operations, and increasingly, our procedures, programs, and planning must recognize this reality.[3]

But the price of interoperability at the national level can be high, and equity can be difficult to achieve. National pride, the importance of each country's military-industrial base, and other economic considerations are part of the picture. Political costs and military risks might result from specific interoperability initiatives, which may lead to decisions not to sell or transfer the most advanced systems and technologies to allies. Risks of proliferation of shared technologies and systems to third parties also enter in: the United States (or its allies) might someday have to fight against its (or their) own systems or

---

[3]See Chairman of the Joint Chiefs of Staff (1996), p. 9.

may find that they have been exploited by hostile states to produce effective countermeasures.

In pursuing interoperability initiatives to support JV 2010 interoperability guidance, one fact is apparent: there are limits on the extent to which any nation is willing to trust another. These limits constrain openness and system interdependencies (e.g., in intelligence, navigation, and communications), which in turn affect interoperability. For example, common or interoperable information systems and databases are vulnerable to disruption, corruption, and theft of data by an expanded number of insiders—a difficult challenge of the information age.

These interoperability problems may be exacerbated if one nation (e.g., the United States) takes actions that are not acceptable to NATO. If for example, the United States wants to defend interests in Southwest Asia (SWA) while NATO strenuously objects, the NATO nations may deny use of their airfields and airspace. This would result in an inability to deploy some combat air forces, a greater expenditure of tanker assets, and possibly the inability to use civilian nonrefuelable airlift.

When there is U.S. and NATO cooperation, overall effectiveness may still take a back seat to national pride, in which case wartime interoperability may become less about maximizing efficiency—in making the whole even better than the sum of its parts—and more about minimizing the burden of politically expedient coalitions. Further, because the United States is largely able to go it alone if necessary, NATO allies may view interoperability efforts as a one-way street, with interoperability compromises and costs unfairly forced on them.[4] Their response might be to accept less well-integrated interoperability levels with the expectation that the United States will ultimately absorb the costs of coalition interoperability shortfalls and inefficiencies.

---

[4]Paradoxically, a U.S. ability to "go it alone" by acting unilaterally may be a requirement for leading a coalition effort insofar as the threat of unilateral action may spur coalition joining by those who hope to influence the objectives and nature of the military action; see Gebhard (1994), p. 39.

## OPERATIONAL AND TACTICAL PERSPECTIVES

Interoperability at the operational and tactical levels is where strategic/political interoperability (discussed above) and technological interoperability (discussed below) come together to help the NATO allies shape the environment, manage crises, and win wars. This is the real-world realm of the warfighter. Interoperability's purpose and focus is to satisfy the political leadership's strategic objectives, within the given constraints and with the maximum possible efficiency and economy of force.

The benefits of interoperability at the operational and tactical levels generally derive from the fungibility or interchangeability of force elements and units. Planning for and conducting NATO-led operations or operations by ad hoc "coalitions of the willing" in out-of-area MTWs or military operations other than war (MOOTWs) involves a process of force "rationalization," i.e., assessing how best to accomplish the mission with the resources available from the coalition members.

The result can vary from a tightly integrated operation (e.g., mixed coalition strike packages) to a coordinated partitioning of the mission or battlespace into separate country-specific chunks. Integration can be achieved through a variety of means, including "interoperable" command centers with standardized communications and computerized data networks, intelligence, surveillance, and reconnaissance (ISR) systems, and force elements, or through ad hoc techniques, procedures, and linkages that include extensive use of liaison officers.

Interoperability-associated costs at the operational and tactical levels tend to result from inefficiencies caused by a number of possible factors outside the immediate control of the warfighters, such as the strategic objectives, strategy and doctrine, role, and systems capabilities of the dominant (whether in the lead or not) coalition partner (for the indefinite future, the United States) and those of the other coalition partners. Coalition operations such as tactical assessments, decisionmaking, planning, force execution, and evaluation are less efficient than U.S. unilateral operations simply because of the number and diversity of the participants. However, unilateral operation may be more costly. Coordination, consensus building, and unanimity lead to delays with potentially costly effects on the operational tempo ("shock and awe") that is the hallmark of U.S. military supremacy. Coalition-related reductions in operational tempo can result in longer conflicts, with resultant increases in material and human costs and possible loss of resolve at the political level.

From the perspective of the dominant partner, these inefficiencies and costs may be exacerbated by the need to divert scarce U.S. military resources to support the partners (e.g., airlift) and to wait for the partners to catch up on mission assignments that could have been more effectively accomplished by U.S. forces. However, these inefficiencies may be necessary to gain political support, access to infrastructure, and use of the airspace of nondominant partners in coalition operations.

Finally, the number and diversity of participants in a coalition increases the chance of military errors, e.g., fratricide or unacceptable collateral damage. These errors must again be hedged against with inherent efficiency penalties.

## TECHNOLOGICAL PERSPECTIVE

This perspective takes our analysis of interoperability into the mechanics of system technical capabilities and interfaces between organizations and systems. It focuses on communications and computers but also involves the technical capabilities of systems and the resulting mission compatibility or incompatibility between the systems (hardware and software) and data of coalition partners.

At the technological level, the benefits of interoperability come primarily from their impacts at the operational and tactical levels in terms of enhancing fungibility and flexibility. (Technology areas include secure voice and data communications, combat identification, and PGWs.)

At the technological interface level, according to the National Institute of Standards and Technology (NIST), the Institute of Electrical and Electronic Engineers (IEEE), the Army Science Board, and the DoD, interoperability is

> The capability of systems to communicate with one another and to exchange and use information including content, format, and semantics.[5]
>
> (1) The ability of two or more systems or components to exchange data and use information;[6] (2) The ability of two or more systems to exchange information and to mutually use the information that are exchanged.[7,8]
>
> The condition achieved among communications-electronic systems or items of communications-electronics equipment when information can be exchanged directly and satisfactorily between them and/or their users. The degree of interoperability should be defined when referring to specific cases.[9]

Recognizing the need for systems to work together adequately in a realistic operational context, many organizations have developed definitions designed to facilitate future investments in systems and to harmonize existing programs. But discussions of such technology-based interoperability initiatives can quickly lose their focus on strategic and operational objectives and become arguments about more tractable tactical and programmatic issues. This phenomenon is not surprising given that technologists have difficulty fully comprehending operational stresses and realities and that operators have difficulty fully understanding the inherent complexity and rigidity of much of today's technology.

In theory, perfectly interoperable systems and data would support the strategic, operational, and tactical interfaces between organizations in ways that are consonant with preexisting agreements on organizations, strategic objectives, and operational concepts. Of course, such perfectly interoperable systems are unlikely to be achieved in practice, and as a result, critical interoperability shortfalls must be identified. Before technological systems are built, two perspectives must therefore be discussed: First, what contribution

---

[5]See National Institute of Standards and Technology (1996).

[6]See IEEE STD 610.12.

[7]These two definitions are quoted in Department of the Army (1997).

[8]See Army Science Board (1995).

[9]See Joint Staff (1999), p. 229.

are such systems expected to make to the organization's strategic objectives? And second, what operational concepts will they enable in future military operations?

Interoperability also derives from the technical capabilities and design of standalone coalition systems. In a coalition with mixed capabilities, interoperability in the sense of seamlessly integrated operations may not be possible. In that case, different technology levels may require that the mission and battlespace be partitioned according to coalition member capabilities, and such assignments may lead to lower warfighting effectiveness and increased costs and risks. Less technologically capable U.S. partners may see such assignments as "second order" or even demeaning relative to U.S. roles and missions.

## SUMMARY

Simply stated, interoperability supports U.S. national security and U.S. national military strategies. It can enable coalition building with coalition partners. It can sustain coalitions by reducing the costs of participation and increasing burden sharing. And it offers an opportunity to enhance future coalition operations. This final benefit confers additional advantages beyond the specific coalition operation. For example, effective allied forces will be better able to carry the continued burden of peace operations while U.S. forces can be redeployed to a major crisis or to an MTW. Furthermore, effective and efficient coalitions will improve the prospects that coalition partners will join future coalitions.

However, the complexity of interoperability and its multiple dimensions make it difficult to understand the nature of the benefits, costs, and tradeoffs that the United States and NATO allies will face in future efforts to improve the interoperability of coalition forces. In fact, much of the value of interoperability is intangible and not easily measured or quantified.

Chapter Three

# INTEROPERABILITY CHALLENGES IN RECENT COALITION OPERATIONS

We reviewed a number of recent coalition operations to identify the challenges that can arise in coalition operations. These challenges provide a starting point for understanding and addressing interoperability in future coalition operations in the new security environment (discussed in Chapter Four) and in the various case studies presented in Chapters Five through Ten.

Our review used the broad concept of interoperability—including strategic, operational, tactical, and technological facets—to gain a clearer understanding of interoperability and interoperability needs in the context of 40 recent U.S. coalition operations that included NATO allies (one of these operations, Allied Force, was a NATO Alliance operation).

## INTEROPERABILITY LESSONS LEARNED

Our review revealed a number of lessons regarding coalitions and interoperability.[1] Because the United States participates in coalitions when undertaking both combat and noncombat operations, interoperability needs to be addressed across the entire spectrum of operations.

---

[1]Appendix A presents supporting data for some of the observations made in this chapter. For more detailed information on all aspects of our review of coalition operations, see Larson et al. (1999).

In addition, because participation by NATO allies in U.S. coalitions varies from situation to situation, interoperability is needed to ensure national-level "plug-and-play." Uncertainty exists in many areas: What missions will be needed? Which countries will participate? Under what conditions will allies join or leave the coalition? What forces will they contribute? Flexible organizational structures, doctrines, procedures, and "open architecture" systems are needed, as are liaison officers to overcome cultural, linguistic, and informational barriers and to facilitate information flows.[2]

Often the United States not only is the single largest contributor to coalition operations but also tends to contribute the broadest range of aircraft, including the capabilities that can provide the C3ISR backbone for the operation. The United States also brings more capable aircraft. For example, recent coalition operations demonstrate the growing divergence between U.S. and NATO allies' air forces in all-weather precision-strike capabilities to minimize collateral damage and employment of standoff weapons, as well as in stealth to minimize the risk of aircraft attrition to enemy defenses.[3]

The strategic, operational, tactical, and technological dimensions of interoperability have problems at all levels. Further, the problems are not confined to the level at which they were observed. Strategic-level interoperability problems, for example, tend to reverberate throughout the operational and tactical levels, as when divergences develop over the political objectives of a military operation, leading to reduced levels of cooperation among coalition members.[4]

Interoperability "workarounds" (short-term and usually incomplete solutions to interoperability problems) as well as longer-term interoperability solutions need to address the fundamental sources of the problem. For example, no amount of operational, tactical, or

---

[2]In the short run, the tools most likely to best manage these frictions are organizational and doctrinal elements that enhance flexibility and adaptiveness and application of experience gained in recent coalition operations, combined with routine exercise and training in a coalition setting.

[3]Comments by senior NATO officials and the new NATO Secretary General George Robertson have highlighted this inequality; see Drozdiak (1999) and Dahlburg (1999).

[4]In a similar vein, the absence of secure communications or inadequate combat identification (an interoperability shortfall at the tactical and technological levels) may greatly increase the risk of aircraft attrition and reverberate up to the strategic level.

technological workarounds can repair an interoperability problem whose origins are fundamentally at the strategic level.[5]

## KEY INTEROPERABILITY CHALLENGES AND WORKAROUNDS

Key interoperability challenges and workarounds at the strategic, operational, tactical, and technological level were also identified. At the strategic level, key interoperability challenges included building and maintaining coalitions (Desert Storm and Allied Force), access restrictions (Desert Thunder/Fox and Deliberate Force), C2 and decisionmaking (Deny Flight and Implementation Force/Stabilization Force [IFOR/SFOR], changing political objectives (Restore/Continue Hope), and the evolving force structure requirements.[6]

At the operational level, force planning, C2, and battle management were among the predominant challenges encountered, followed by information exchange and security issues. In addition, we found that nations, including the United States, are likely to continue to maintain direct national control of their national and theater ISR assets rather than contribute them to a larger, shared pool under direct coalition control.

At the tactical level, the key interoperability challenges encountered in the coalition operations were diverse. In three cases (Desert Storm, Restore/Continue Hope, Deliberate Force), the particulars differed, but a key challenge was integrating coalition forces of varying performance capabilities into tactical operations. In Desert Storm and Deliberate Force, problems with coalition tactical communications and combat identification led to division of the battlespace to separate (and deconflict) air and ground coalition forces; the United States carried the greatest burden for some missions (e.g., precision strike). By contrast, in Restore/Continue Hope, a key tacti-

---

[5]A good example is Somalia, in which a lack of unity of purpose compromised unity of effort and command and led to a chain of command that proved incapable of preventing or mitigating the consequences of downed helicopters. By contrast, had there been consensus at the higher (e.g., strategic and operational) levels, these lower-level interoperability problems would have been less likely and more manageable.

[6]See Appendix A for a brief description of the specific operations discussed in this chapter.

cal issue was a shortfall in coalition C3 capabilities; the workaround in this case was provision of communications assets by the United States and extensive use of liaison officers.

At the technological level, the lack of automated tools and compatible and secure communication systems made it difficult to build and disseminate the air tasking order (ATO) or its equivalent and to establish and maintain secure communications among coalition aircraft. This was a key challenge in Desert Storm, Deny Flight, Restore/Continue Hope, and Allied Force. Workarounds to address these ATO-related challenges included manual processes and physical dissemination. Workarounds to address the lack of adequate secure communication systems included use of unsecure communications and, when possible, use of codes, taking the associated risk of information compromise.

## BROADER LESSONS FOR INTEROPERABILITY PLANNING

Our review also revealed other, broader lessons for interoperability planning. For example, we found that even when coalition partners agree on an overall objective and military mission, they may diverge about how to accomplish that objective or about the amount of risk they are willing to assume. In the worst case, agreement may be somewhat nominal—representing a papering over rather than a resolution of differences. When political motives are misaligned, no amount of interoperability, technological or otherwise, can mitigate the problem.

A related lesson concerns commanders and political leaders who may face challenges in balancing each nation's political needs against the military requirements of the operation. This is particularly important when political guidance changes in the course of an operation. Such tensions can complicate both C2 (the vertical dimension) and coordination (the horizontal dimension).

Finally, anecdotal historical evidence suggests that it is not unreasonable to view policy leadership as a function of willingness to accept (or share) risk. In this view, the more risk that the United States is willing to accept, the stronger its negotiating position will be within the coalition. In cases where the stakes for the United States are low, the willingness to accept risks may also be commensurately low and U.S. ability to manage the coalition problematic. In such cases (e.g., Somalia), the United States may face difficulties in forging a common purpose, a common effort, and a harmonized chain of command; in instilling coalition discipline; and in preventing defections or subversion of coalition aims.

At the other extreme, when the stakes are sufficiently high that the United States indicates a willingness to go it alone and accept most or all of the risks, coalition partners may be able to influence only U.S. aims, conduct, and management of the coalition. Anecdotal historical support for this proposition can be found in Desert Storm, Deliberate Force, and Desert Fox, where U.S. willingness to commit the greatest share of forces and fly the most challenging missions arguably strengthened its role in running the air campaign.

Chapter Four

# NEW TRENDS THAT MAY AFFECT FUTURE INTEROPERABILITY

In this chapter, we examine new technology, security, and military trends—as reflected in top-level DoD guidance and visions of future military operations—to identify interoperability challenges that U.S. and NATO allies' air forces will need to address over the next decade. We consider the following factors: the changing international security environment, the budgetary and programmatic environment, and the potentially widening gap between U.S. and NATO allies' military capabilities.

## THE INTERNATIONAL SECURITY ENVIRONMENT

Future interoperability needs must be understood in the context of recent changes in the international security environment that affect coalition operations. Of particular interest are changes in NATO's security environment that have led to new missions for NATO, and the United States' increasing reliance on European coalitions and organizations.

Arguably, the international security environment is more stressing now than it was during the Cold War. This situation results from potentially fast-changing circumstances, the wider range of possible contingencies to which forces must respond, and the interface between forces of different nations at different levels. Thus, interoperability becomes increasingly important because of the new missions likely to develop in this environment, the increasing U.S. reliance on European "coalitions of the willing," and a political environment that is less tolerant of mistakes leading to "unnecessary" casualties or collateral damage.

U.S. national security strategy increasingly emphasizes the role of Europe in coalition operations and reliance on European allies to achieve global security objectives. NATO's security environment has also changed: NATO continues to reengineer itself to undertake a range of new missions, and its focus is shifting from a strategic concept based on the threat of Soviet aggression to one that will improve NATO's ability to manage internal instabilities on the periphery and in out-of-area operations.

## Changes in NATO's Security Environment

Traditionally, threat-based analyses (What are the threats? How can they best be countered?) have provided the overall context and justification for establishing military needs and for developing new concepts and systems to address the identified needs. Since the end of the Cold War, however, the international security environment—and U.S. military operations—have been dominated by less predictable events such as civil wars and regional crises. There is little reason to believe that this situation will change in the immediate future.

SWA, at least until Saddam passes, will likely continue to be a focus of U.S. and, to a lesser extent, European concern.[1] In the longer term, there is the possibility of a revanchist Russia or an emergent China as a "regional," "niche," or even "peer" competitor (although the level of European interest in China is less obvious than that in Russia).

Recurring crises below the MTW level[2] suggest that proliferation of ballistic missiles and weapons of mass destruction (WMD) may be an area of increasing concern, leading to the deployment of air defense

---

[1]Although not as obviously an area of interest to Europe, North Korea may also be a future locale for combined U.S.-European action. Accordingly, interoperability issues can be expected to center on South Korea and, to a lesser extent, Japan.

[2]Examples include the crisis with Iraq over United Nations Special Commission (UNSCOM) inspections, the October 1994 crisis with North Korea over nuclear weapons, the September 1998 launch of a North Korean theater ballistic missile (TBM) over Japan, and the deployment of Patriot missiles to Israel in late 1998.

and C3ISR capabilities. And, as demonstrated by the U.S. strike on Libya in 1986 and the strike on Bin Laden's facilities, counterterrorist strikes aimed at preventing further acts of terrorism are also likely to continue to be part of U.S. future operations.

U.S. and multinational involvement in Somalia and Bosnia, as well as events in Kosovo and elsewhere, suggest that in some cases internal conflicts can create requirements for the use of force. Peace operations in Bosnia, Kosovo, and other places can be expected to continue to create demands on the United States and its European allies.[3]

There is little reason to believe that the prevalence of often complex and sometimes risky humanitarian disasters in Africa will end any time soon. These situations will likely continue to require attention and some level of interoperability, even when U.S. combat forces are not participating.[4] We also note that some allies are more worried about U.S. unilateralism than about some of the threats discussed above.

In such operations, the United States and NATO allies may also encounter new, more challenging threats such as advanced mobile surfaces-to-air missiles (SAMs), tactical cruise and ballistic missiles (perhaps low-observable variants or those carrying WMD), and more mobile force elements (mobile C2 nodes and dismounted forces). Along with changes in the nature of warfare (e.g., nonlinear as opposed to linear battlefields), and with increased employment of more advanced capabilities such as information warfare, these developments may present new interoperability problems.

---

[3]The Petersburg Declaration of 1992 foresees that European states in the Western European Union (WEU) could undertake a range of military actions, including humanitarian and rescue operations, peacekeeping, and the use of combat forces in crisis management, including peacemaking. The European Union's Treaty of Amsterdam also envisions a military role for WEU states.

[4]This is true even if interoperability considerations center on issues such as compatibility in handling outsize cargo. In addition to frequent participation in such operations by the United States and its NATO allies, increasing attention is being paid to promoting the development of African nations' own capabilities to carry more of the burden of many of these operations. This suggests another area of potential interoperability opportunities or challenges.

By all accounts, the post–Cold War world has left NATO and its member nations with a less compelling set of security problems than those posed by the Soviet Union and the Warsaw Pact. National survival is no longer at stake as it was in the Cold War. Accordingly, NATO and its member nations have had to adapt their capabilities and organizations to address a different set of challenges, including conflicts such as Bosnia, and to tailor their planning to address those challenges. NATO has also tried to tie itself more closely to other political and security institutions that are relevant to European security, has widened the circle of nations that participate, and has thereby added additional degrees of freedom to interoperability requirements.[5] For example, the Partnership for Peace (PfP) includes the 19 NATO nations as well as 24 others.[6]

In the face of less compelling threats, the importance of minimizing casualties—including those of friends and even possibly adversaries—has arguably increased in the post–Cold War world. This is because NATO politicians who ultimately decided if military intervention is warranted put a high value on minimizing casualties in efforts to mitigate public opposition. Thus, any given intervention will likely be judged by the electorate and is likely to be undertaken only if casualties are expected to be commensurate with the importance of the interests and values that are engaged.[7]

In a similar vein, by reducing fratricide within NATO, the NATO nations can reduce potential frictions with each other's publics. The implications for interoperability are numerous and include comparable (i.e., easily substituted) all-weather precision-strike capabilities across NATO allies' air forces, improved secure communications and combat identification, and other information-intensive capabilities. These factors suggest that research to identify interoperability

---

[5]Among these institutions are the WEU, which, until the Anglo-French agreement of December 1998, was slated to form the core of the prospective European Security and Defence Identity (ESDI); the Euro-Atlantic Partnership Council (EAPC); and the PfP—all of which include non-NATO members.

[6]Non-NATO members of the PfP are Austria, Finland, Sweden, Bulgaria, Estonia, Latvia, Lithuania, Romania, Slovakia, Albania, Armenia, Azerbaijan, Belarus, the Former Yugoslav Republic of Macedonia (FYROM), Georgia, Kazakhstan, Kygyzstan, Moldova, Russia, Slovenia, Turkmenistan, Ukraine, Uzbekistan, and Malta.

[7]This is also true, generally to a lesser degree, in the case of noncombatant casualties. See Mueller (1994) and Larson (1996).

needs and longer-term solutions should focus on the capabilities and levels of interoperability that will be needed to perform future high-interest missions at acceptable performance levels.

### New Missions for NATO

Since the collapse of the Soviet Union and the demise of the Warsaw Pact, challenges to security in Europe have derived more from instability arising within countries than from external threats. This poses a requirement for fighting highly limited wars that are infused with political constraints.

The emergence of these missions as a key focus for planning and action by NATO and its member nations has challenged the adequacy of extant doctrine, organizations, training, exercises, and systems in ways that were never envisioned in planning to deter and, if need be, repel a Warsaw Pact attack. The constraints imposed on these new missions—on rules of engagement, civilian and military casualties (including fratricide), and the like—pose unique challenges for interoperability in coalition and Alliance operations. Recognizing these new challenges, NATO is reengineering itself to undertake a range of out-of-NATO-area power projection missions.[8]

## THE BUDGETARY AND PROGRAMMATIC ENVIRONMENT

An equally challenging budgetary and programmatic environment is emerging in which interoperability enhancements are seen as a means of achieving efficiency and ensuring that critical gaps between the United States and its NATO allies can be minimized.

### Tighter Defense Budgets

As a result of the diminished threat environment, U.S. and NATO allies' defense budgets have declined. This decline has been disproportionately severe in the investment account, with a few large programs consuming most of that budget. As a consequence, whatever efforts are made to achieve interoperability will need to fit well

---

[8]The C2 case study presented in Chapter Five discusses NATO reengineering efforts.

within constrained resources. Further, in pursuing interoperability, each nation may be asked to trade off national military capability for interoperability with other NATO members. The extent to which nations are willing to sacrifice military capability to achieve interoperability is a critical consideration.

## The Political and Economic Aspects of Defense Consolidation

Interoperability initiatives offer a number of approaches for promoting the further rationalization of alliance-wide defense industries—for example, through collaborative efforts (e.g., EF-2000, European Joint Strike Fighter) or through single-source efforts (e.g., U.S. manufacture and sale of F-16s). Because these efforts frequently result in smaller pies being divided up among a smaller number of commercial players, pressures for rationalization and consolidation are in tension with continued national desires to preserve the perceived economic benefits of national defense industries (e.g., revenues and employment). Political-economic interests in many quarters (including the United States) are likely to press for equity over economic efficiency and may impede otherwise promising interoperability initiatives.

This dynamic budgetary and programmatic environment makes the question of performance and interoperability gaps a complex one. Further complicating the picture, gaps between the capabilities of the United States and its key allies could emerge or widen as a result of the United States' JV 2010 capabilities and the sorts of top-level future operational concepts and emerging systems described later in this report.

## Concurrent Development and Introduction of New NATO Capabilities

A related concern can be found in the research, development, and acquisition activities of the United States' NATO allies. These development programs may be proceeding with inadequate consideration of the interoperability requirements for operating in a coalition with the United States, potentially producing a widening gap in performance and interoperability.[9] The NATO allies have a number of air and C3ISR systems that are well along in research, development, or acquisition—e.g., EF-2000 (Typhoon), Rafale, and the Airborne Stand-Off Radar (ASTOR).[10] Although planned to be NATO-interoperable, these systems will need to be integrated with U.S. capabilities at some investment cost to achieve maximum benefit. Furthermore, a strong argument can be made that interoperability should be addressed early in the design, development, and acquisition process so that least-cost, longer-term solutions can be found for integrating capabilities into effective and efficient combined operations. The alternative—attempting to integrate deployed systems—means that integration is likely to amount to little more than improvised workarounds that are less effective than systematic integration of elements in a larger system.

In summary, in the near term, U.S.-NATO interoperability may be limited by the United States' and NATO allies' piecemeal introduction of new systems, standards, doctrine, and organizations. The dynamic acquisition environment will pose challenges to the integration and interoperability of these new capabilities with operational concepts, doctrine, and organizations.

## A POTENTIALLY WIDENING GAP IN U.S.-NATO CAPABILITIES

With JV 2010, the U.S. military has embarked on an ambitious path that may mean that the gap between U.S. capabilities and those of its adversaries will widen further as the United States capitalizes on

---

[9]Development and procurement of common systems by the NATO nations can certainly foster interoperability. The NATO AWACS program and the MIDS terminal discussed in Chapter Seven and Nine, respectively, are examples. Today this is more problematic given the political and economic stakes, including the desire to nurture emerging high-technology industries and to ensure sizable work shares. In addition, international programs have generally exhibited greater cost growth and schedule slippage than national development efforts. For a more complete discussion of this last point, see Lorell and Lowell (1995).

[10]The International Institute for Strategic Studies' *Military Balance* provides a comprehensive listing of NATO members' current aircraft, C3ISR, and other acquisition programs.

technological prowess in information superiority, stealth, standoff, precision, joint interoperability, and other capabilities.[11]

The gap between the capabilities of the United States and its key allies may also be widening to the extent that the NATO allies may not be able to perform military missions at U.S. performance levels.[12] Without a compelling threat, if NATO and its member nations' capabilities and operational concepts become outdated or incompatible with those of the United States, NATO allies' participation in coalition operations may become increasingly marginal and could ultimately erode the Alliance. If a compelling threat should emerge, the result would be a weakened NATO capability to respond. In short, the stakes of lack of interoperability are high.

If there are near-term challenges to the interoperability of U.S. and NATO allies' air forces, the far-term challenges may be even more sobering. At its most fundamental level, JV 2010 represents an objective design point for future U.S. military forces, doctrine, organization, training, and equipment. Thus, unless NATO—selectively or as a whole—moves toward a parallel or complementary design point, interoperability may become an increasingly difficult problem.[13]

Because NATO allies may ultimately need to interoperate with U.S. JV 2010 forces, it is essential, as a first step, to describe JV 2010 in tangible terms to reveal potential future interoperability needs. We describe JV 2010 below in terms of top-level operational concepts, along with the emerging systems, standards, doctrine, organization,

---

[11]The United States has a margin in many capabilities, including stealth, standoff and cruise missiles, and information.

[12]Indeed, the U.S. focus on the effectiveness and efficiency of military operations may not be shared by its NATO allies, some of whom may place a higher premium on consensus and equity.

[13]An issue for further study centers on the best U.S. strategies for achieving interoperability with NATO allies. An alternative to NATO-wide interoperability efforts would be to pursue bilateral or multilateral interoperability initiatives with selected or key allies. For example, one strategy would be to invest heavily in improving interoperability with the largest allies, including the United Kingdom, Germany, France, or Italy, which have relatively robust air capabilities. Another strategy would be to focus on smaller allies such as Denmark and Norway, which are more inclined to buy U.S. equipment and will need to integrate with a larger ally such as the United States to have access to the full suite of air and C3ISR capabilities.

and training and exercise options that may provide the enabling capabilities for performing these missions.

Top-level DoD guidance such as JV 2010 and Air Force *Global Engagement* postulate new military operational concepts: precision engagement, dominant maneuver, focused logistics, and full-dimensional protection.[14] These concepts are enabled by improved C2 and intelligence assured by information superiority, which is in turn made possible by the dramatic advances in information technologies (e.g., navigation, guidance, computers, and communications) and new ways of doing business that rely on off-the-shelf technologies and commercial standards and solutions.

The major improvements in C3ISR capabilities envisioned in JV 2010 may enable new operational concepts and activities. Improvements in information and systems integration technologies will also affect future military operations by providing decisionmakers with accurate and timely information. Information technology will improve the ability to see, prioritize, assign, and assess information. The fusion of all-source intelligence with the fluid integration of sensors, platforms, command organizations, and logistic support centers will allow a greater number of operational tasks to be accomplished faster.

Advances in computer processing, precise global positioning, sensor technologies, and telecommunications will allow for the accurate determination of locations of friendly and enemy forces as well as the ability to collect, process, and distribute relevant data to thousands of locations. Forces harnessing the capabilities potentially available from this "system of systems" will gain dominant battlespace awareness—an interactive "picture" that will yield much more accurate assessments of friendly and enemy operations. Although this interactive picture will not eliminate the fog of war, dominant battlespace awareness will improve situational awareness, decrease response time, and make the battlespace considerably more transparent to those who achieve it.[15]

---

[14]See Appendix B for definitions of these new operational concepts. See also Chairman of the Joint Chiefs of Staff (1996).

[15]Ibid.

We next describe three overarching concepts that characterize the contribution of C3ISR assets to future force-level and unit-level air operations: precision strike, network-centric collaborative operations, and dynamic assessment, planning, and execution (DAPE).

## Precision Strike

Precision strike operations are the precise application of weapons against critical points of individual targets or nodes of target sets to achieve damage with increased efficiency and minimal collateral damage. The development and fielding of a large number and variety of all-weather PGWs, and the increasing availability of precise information provided by ISR assets are the key enablers for such operations. U.S. investment in such capabilities proved their value in recent coalition operations. NATO allies have recognized the value of such systems but have yet to make comparable investments.

## Network-Centric Collaborative Force-Level and Unit-Level Operations

A network-centric collaborative environment for future force-level and unit-level assessment, planning, and execution will ensure unity of effort (attainment of commanders' objectives) while providing all relevant parties easy access to the right data, at the right time, in proper format, at the right locations, and at the right security level. In this concept, access to information would be provided on a global grid consisting of terrestrial, airborne, and space connectivity assets. Information for the grid would be provided by a wide range of sensors balanced with data analysis resources. Unity of effort and maintenance of control are to be achieved through new network protocols (policy, semantics, and procedures) that tie together sensor collection, data analysis resources, and decisionmakers (e.g., planners, controllers, and shooters).

Five key prerequisites for such a distributed collaborative environment are (1) sensors capable of collecting (night and day and in poor and good weather) accurate and sufficient data on a wide range of stationary and mobile targets across the battlefield, (2) sufficient analytical resources to exploit collected data in a timely fashion, (3) enhanced decision aids, (4) robust multilevel security communications, and (5) trained personnel who can operate confidently in such an environment. To accomplish this, major impediments to information sharing and the current predilection for collocated operations (with face-to-face contact) as opposed to distributed operations will have to be overcome.

## Dynamic Assessment, Planning, and Execution

Implicit in JV 2010 is the DAPE concept. The traditional 72-hour ATO process will have to be modified to allow for the retasking of substantial numbers of air missions to address time-critical targets. A necessary enabler for DAPE is timely and accurate situation awareness of adversary forces, U.S. and coalition partner forces, and neutrals. A common operational picture (COP), a common tactical picture (CTP), and a single integrated air picture (SIAP) will provide operators with data needed for situational awareness. However, full use of these data requires evolution in doctrine and tactics, new weapon systems, C2 decision aids, and operators trained to perform in such an environment. There must also be recognition that traditional, deliberate ATO planning will still take place. Such capabilities are currently under development.

These top-level concepts of JV 2010 are likely to reveal—or produce—additional interoperability gaps between U.S. and NATO allies' air forces and, if left unattended, will lead to even wider gaps in capabilities.

## Defence Capabilities Initiative

Recognizing that these and other shortfalls exist among its members, NATO endorsed the Defence Capabilities Initiative in April 1999 to meet the challenges of the present and foreseeable security environment[16] The most important areas identified for improvement were the deployability and mobility of Alliance forces, their sustainability and logistics, their survivability and effective engagement capability, and the necessary C2 and information

[16]See Solana (1999).

systems.[17] Note that these improvements are needed not only for future NATO air operations but also for other future coalition air operations in which NATO members are likely to participate.

Both the former and the current NATO Secretary General have indicated that addressing these shortfalls will require increased defense expenditures by the NATO allies:

> It's a matter of political will and harmonizing Europe's military industries, but most of all it's a matter of money. It's hard to say just how much will be enough. Defense budgets will have to rise, but we could accomplish a lot just through better coordination of the way we spend our money.[18]
>
> That [getting relevant capabilities for the future] means we've got to reorder spending priorities and, in a lot of countries, spend more on defense if we're going to have the investment in security for the future that the continent needs.[19]

With current budgetary constraints and weak public support in some countries for defense expenditures, it is not clear that the NATO allies will make the necessary investments by increasing the defense budget or by shifting resources from personnel and operations and maintenance to investment to acquire the needed capabilities.[20] According to Secretary of Defense William Cohen: The challenge Europe faces today is to turn words into action.[21]

---

[17]See NATO (1999b).

[18]Javier Solana as quoted in Drozdiak (1999).

[19]George Robertson as quoted in Dahlburg (1999).

[20]According to the U.S. General Accounting Office, NATO countries have made some progress since 1991 to increase the mobility and deployability of their forces to conduct out-of-area offensive campaigns, but "the alliance still faces challenges to continue to improve mobility and deployability capabilities" (GAO, 1999, p. 6). During the Cold War, this capability was not needed, as the countries were planning to fight in place with logistical support provided by fixed facilities.

[21]See Cohen (1999).

## CASE STUDIES

In the next six chapters, we examine key ongoing U.S. and allied programs that have major implications for future interoperability, from the strategic down to the technological level, and suggest actions the U.S. Air Force can take to address interoperability challenges that the U.S. and NATO allies' air forces will face over the next decade. Mindful of the current budgetary environment on both sides of the Atlantic, we emphasize lower-cost short- and medium-term solution directions (e.g., actions regarding organizations, doctrine, standard setting, and systems based on available information technology rather than new, major weapon programs)[22] that will encourage the United States' NATO allies to "turn words into action."

---

[22]This does not imply that efforts such as NATO's Alliance Ground Surveillance capability and the Joint Strike Fighter should be abandoned but rather that a common platform approach should not be the dominant factor in addressing interoperability challenges.

Chapter Five

# COMMAND AND CONTROL

The diminution of the Soviet strategic threat in Europe has led to a significant evolution of NATO strategic focus from the defense of NATO member territories[1] to one that now includes out-of-NATO-area operations in missions such as peacekeeping, crisis response, and crisis management. Before the recent NATO reorganization, three regional commanders-in-chief (CINCs) commanded and controlled air operations in their sector of responsibility in support of Article 5 operations. This will not be the case in out-of-NATO-area operations, as both the areas of responsibility and the command structure will be different.

This shift in focus has had and will continue to have significant implications relating to how NATO nations' air forces are commanded and controlled. The United States—as a member of NATO and as an interested party in coalition operations with NATO members—must consider and effectively address the changes in C2 doctrine, organization, procedures, systems, and personnel that have arisen from this strategic shift.

At the NATO Summit held in Washington, D.C., on April 23–24, 1999, NATO released an updated Strategic Concept consistent with the

---

[1]The authority for such actions was based on Article 5 of *The North Atlantic Treaty* (NATO, 1998), which states that "the Parties agree that an armed attack against one or more of them in Europe or North America shall be considered an attack against them all," and on Article 51 of the *Charter of the United Nations* (United Nations, 1945), which recognizes the "inherent right of individual or collective self-defence if an armed attack occurs against a Member of the United Nations."

new security environment.[2] This document describes the new security environment, specifies the Alliance's approach to security, and provides guidelines for adaptation of its military forces, particularly NATO's C2 structure:

> NATO's command structure will be able to undertake command and control of the full range of the Alliance's military missions including the use of deployable combined and joint HQs [headquarters], in particular CJTF [Combined Joint Task Force] headquarters, to command and control multinational and multiservice forces. It will also be able to support operations under the political and strategic direction either of the WEU or as otherwise agreed, thereby contributing to the development of the ESDI within the Alliance, and to conduct NATO-led non–Article 5 crisis response operations in which Partners and other countries may participate.[3]

The case study discussed in this chapter examines elements of U.S. and NATO C2 strategic and operational structures, air campaign planning and execution practices and procedures, force-level planning systems, and information-sharing arrangements in the context of recent NATO reorganization and future out-of-area coalition operations.

We focus on interoperability between U.S. and NATO capabilities because only a few NATO countries have extensive national air C2 capabilities: most NATO allies rely on NATO-wide capabilities and rationalize their spending by investing in such capabilities. Therefore, if they participate in non-NATO alliance coalition operations with the United States, they will likely bring NATO C2 capabilities (practices, procedures, systems, and personnel). If these capabilities differ substantially from those of the United States, interoperability problems will arise that did not previously exist.

This case study highlights key differences and potential synergies between U.S. and NATO practices, procedures, and associated systems. It also emphasizes the importance of thinking about interop-

---

[2]See NATO (1999a).

[3]Ibid (para. 53c).

erability challenges from the strategic level to the technological level and the relationships between these levels.[4]

## STRATEGIC LEVEL

The recently approved changes in command structure have important implications for the future of NATO. Although the Supreme Allied Commander Europe (SACEUR) remains a U.S. officer, other command structure changes reflect a somewhat more European emphasis. The number of Regional Commands (RCs) has been reduced from three (North, Central, and South) to two (North and South) (see Figure 5.1). One of these commanders (at RC NORTH) has become a European billet, rotating between a British and a German officer. The RC SOUTH commander remains a U.S. flag officer.

Accompanying these changes in the top-level command structure is the concept of Joint Sub-Regional Commands (JSRCs), whose commanders are subordinate to the regional commanders. The JSRCs are army officers of seven host nations (Germany, Italy, Spain, Norway, Denmark, Turkey, Greece) who maintain a regional presence for the purposes of air sovereignty, infrastructure maintenance, and the like. These commanders also have responsibilities for accommodating an influx of forces during wartime. Therefore, the regional infrastructure, systems, and practices and procedures in place at a JSRC are likely to have an impact (at least initially) if forces from elsewhere are introduced to augment the existing presence.

Accompanying the JSRCs is a set of Combined Air Operations Centers (CAOCs). Under the NATO reorganization, there are nine "static" CAOCs (in Germany, Italy, the United Kingdom, Norway, Denmark, Spain, Portugal, Turkey, and Greece), plus two deployable CAOCs (DCAOCs) garrisoned in Germany (the exact location is not

---

[4]Because of the hierarchical nature of C2, we found it useful to follow the multilevel construct (strategic, operational, tactical, technological) for presenting the results of our interoperability analysis. Although the importance of thinking about interoperability from the strategic to the technological is important for the other case studies as well, we found that construct to be awkward for presenting the analysis because it is difficult to separate the results into each of the levels. In several case studies, we found that a historical or programmatic approach was more useful.

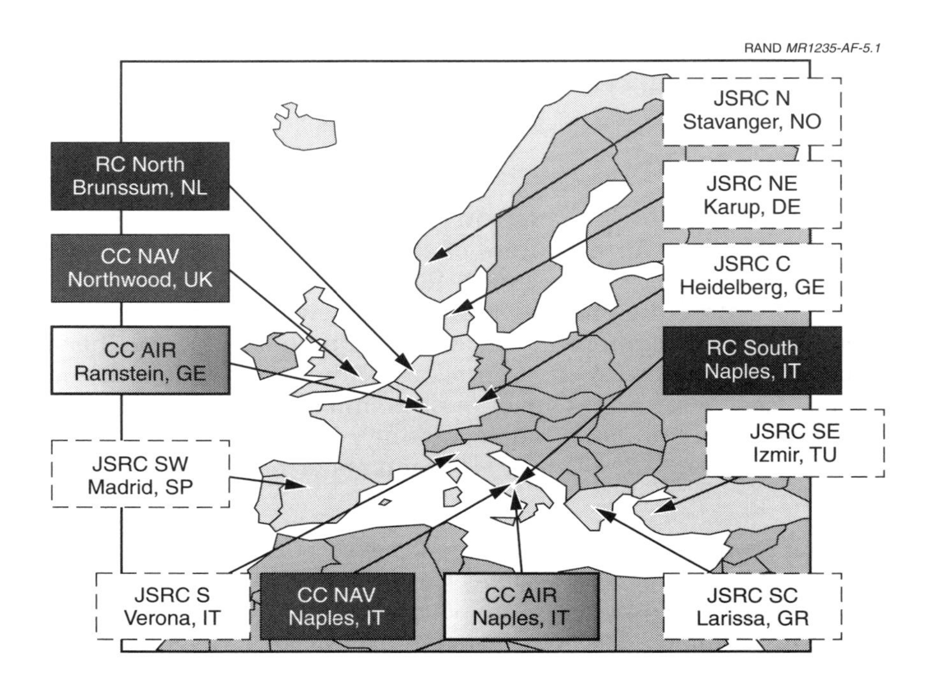


**Figure 5.1—New NATO Regional and Subregional Command Structure**

finalized) and Italy (see Figure 5.2). As with the JSRC command posts, host nation "ownership" is important for two reasons. First, these CAOCs serve as a starting point for augmentation from the NATO air component commanders (CC AIR) in Germany and Italy, as well as national forces. Second, during peacetime the CAOCs also fulfill certain national functions, such as air sovereignty, for host nations. The distribution of CAOCs and JSRC posts is a result of both external, threat-related factors and political considerations within NATO.

## Current Command Constructs

Four basic types of command arrangements can be envisioned in coalition operations with NATO allies. The most common operations in recent history were not conducted within the NATO com-

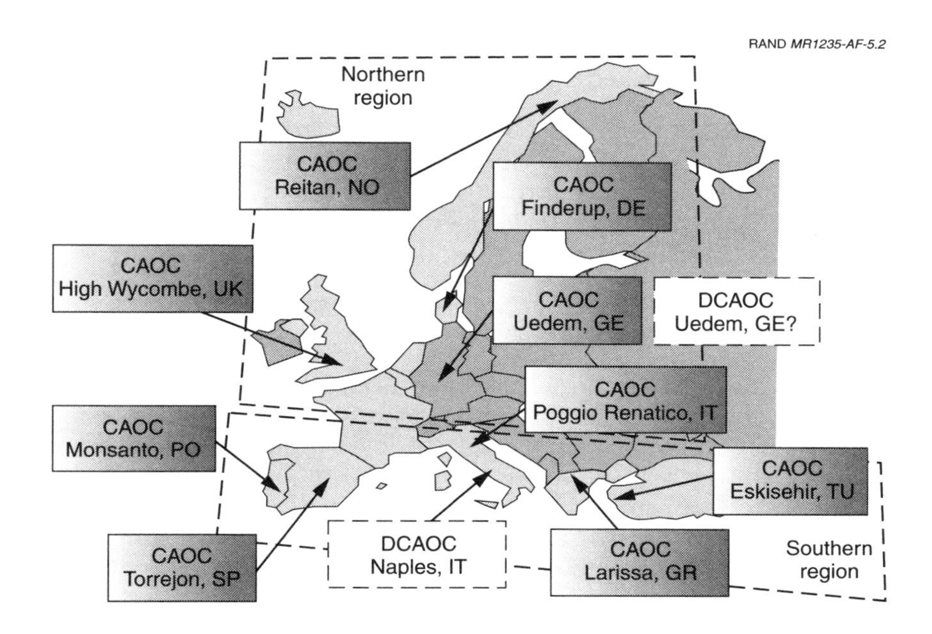


**Figure 5.2—New NATO Combined Air Operations Centers**

mand structure; many operations have been led by the United States with the participation of a limited subset of NATO allies and other coalition members.

In such cases, the United States will likely bring the greatest assets. Because the preponderance of political will, C2 assets, and combat forces will likely come from the United States, it is also likely that U.S. practices and procedures will be followed. However, this reasoning does not underestimate the key contributions of allies in terms of overflight rights, basing, and other support infrastructure, which may provide great leverage depending on the exact circumstances of the conflict.

The most basic type of command arrangement within NATO is the traditional Article 5 operation in defense of NATO. This type of operation has been practiced for years, and as pointed out earlier, the institutions and practices and procedures of NATO itself have been constructed for this type of contingency. For these reasons, and given that the European allies will be the largest stakeholders in this case, NATO systems and doctrine are likely to dominate.

Both the U.S.-led coalition construct and the Article 5 construct represent relatively well-known situations. However, two other types of command arrangements currently evolving in NATO represent important changes at the strategic, operational, tactical, and technological levels. These two constructs will be examined in turn.

## Evolving Command Constructs

An evolving CJTF doctrine has been drafted that has significant implications for the conduct of power projection and out-of-area operations. From the operational level downward, the CJTF doctrine appears on the surface to be very similar to U.S. CJTF doctrine. However, important differences exist at the highest level of command relationships. In the event of an out-of-area contingency, SACEUR is authorized to designate a CJTF commander (ComCJTF) from anywhere within the NATO command structure. Thus, a regional commander (North or South) may or may not be chosen to be the ComCJTF. For example, a different European flag officer may be chosen owing to regional familiarity or relevant service knowledge (air, maritime, land). Air, maritime, and land component commanders are selected to support the ComCJTF; they will provide the CJTF with forces from the standing NATO component commands (CC AIR, CC NAV, and JSRCs) and from supplemental national sources. One of the DCAOCs would likely be deployed to support the Combined Joint Force Air Component Commander (CJFACC) by augmenting the battle staff of the appropriate CAOC.

Although doctrinally similar in many ways to the U.S. CJTF concept, the new CJTF doctrine has two important distinguishing features. First, the appointment of a ComCJTF who is not a regional commander has the potential to represent a deviation from existing practices and procedures used within the standing peacetime command structure. Second, the authority and responsibility with respect to SACEUR, the North Atlantic Council (NAC), and national command authorities remains unclear. These points are discussed in the next section.

The second emerging command construct is known as the joint operational area (JOA) construct. In the case of an emerging regional conflict near a NATO border, a JSRC may request support in from the air, maritime, and land component commanders. As in the case of the CJTF construct, the question of practices and procedures at the level of the JSRC and supporting static CAOC becomes an issue.

Both the CJTF and JOA command arrangements represent potential sources of divergence from established U.S. practices and procedures, but they should also be viewed as targets of opportunity to exert influence to the degree to which they remain uncertain and are still evolving.[5] The lack of constructive U.S. participation within organizations responsible for making changes at this level can have a substantial, cascading impact at the operational, tactical, and technological levels. The nature of these impacts will be described in the following sections.

## OPERATIONAL AND TACTICAL LEVELS

At the operational and tactical levels, discrepancies in policies, practices, and procedures place constraints on the maximum level of interoperability that can be attained between forces. The lack of top-level consensus on strategic and operational objectives can also constrain the ability to work together effectively. As we have defined interoperability, effectiveness is the key metric of interoperability. If discrepancies in force mix, practices, and procedures between the United States and NATO are taken as a given, every effort should be made to provide workarounds so that negative impacts on effectiveness are minimized.

The enduring principle of C2 in the United States has been "centralized control with decentralized execution." To the extent that this principle is threatened, the efficiency and perhaps effectiveness of coalition operations are also threatened. The lack of top-level consensus on objectives can have a cascading impact on

---

[5]We are suggesting the adoption of U.S. warfighting practices and procedures for two reasons: (1) in the near future, the preponderance of C2 assets and combat forces will likely come from the United States, and (2) the United States has developed and established CJTF and air operations center (AOC) practices and procedures that have proven effective over time for a variety of operations.

the effectiveness of a campaign. A campaign plan with clear, well-articulated objectives and strategy is essential to maximizing chances of a successful outcome. If these conditions are not met, a number of possible consequences may result.

First, because of political sensitivities, the delegation of decisionmaking authority to subordinate commanders may not occur. Information technology increasingly allows top-level decisionmakers to observe detailed aspects of combat in near-real-time. It also allows force-level decisionmakers to be easily distracted by tactical issues and to concentrate less on operational issues.

Second, the sharing of tactical information—such as targeting data, ATO information, and airspace control order (ACO) information—that normally happens as a matter of course may become a matter for negotiation at higher echelons of command. Alternatively, information sharing may have to occur at lower echelons on an ad hoc basis to prevent highly undesirable results, such as fratricide. Workarounds that result from these circumstances may not be entirely adequate or uniform across the force.

Most workarounds are, by definition, not matters of policy but instead result from interactions between individuals or small groups who are familiar with each other. The trust and respect essential to accommodating these workarounds may be gained only over time. For example, procedural workarounds between airborne controllers and fighter pilots can occur only if these groups interact with each other face to face, which implies that they are based from the same location and interact on a daily basis. If the United States intends to pursue network-centric warfare, U.S. principles will need to be shaped by these personal considerations, which are part of peacetime and coalition operations.

Operational issues include not only activities at the CAOC, but the activities of crucial air and ground control and surveillance assets as well. In the case of AWACS, U.S. doctrine has incorporated the notion of air control (in addition to surveillance) for some time. In NATO, however, NATO Airborne Early Warning Force Command personnel have only recently begun to perform this function, and new NATO aircraft retrofits will include more consoles that can potentially be used as weapon control stations.

The operational doctrine and training of U.S. and NATO AWACS (and U.K. and French AWACS) should be harmonized to decrease the burden on personnel, since shortfalls are difficult to accommodate with augmentees on a fixed-size aircraft. At a minimum, the capabilities of these aircraft should continue to be harmonized to ensure that the different platforms are fungible at the level of functionality. With respect to ground surveillance and control, the same principles apply. These areas will be discussed in subsequent case studies on airborne surveillance and control (see Chapters Seven and Eight).

At the tactical level, rules of engagement (ROEs) and weapon systems will likely differ substantially between the United States and NATO. While the United States is rapidly acquiring standoff and PGWs, many NATO allies are moving more slowly or are buying smaller quantities of munitions. These differences also affect ROEs, with many European nations requiring eyes-on-target. Different mixes of fighters also help determine under what circumstances different capabilities (fast and low versus high and slow) should be allocated at the force level. In many cases, even similar weapon systems with different munitions (say, F-16s with precision versus nonprecision weapons) are not fungible and may be employed only in a certain threat environment or under certain weather conditions. These factors complicate the force-level planning process. This subject will be discussed in the case study on fighters and weapons (see Chapter Ten).

## TECHNOLOGICAL LEVEL

In this section, we concentrate on force-level planning and execution monitoring and on information exchange. As we have seen, information exchange is a broad topic that encompasses issues at different levels, from intelligence-sharing policies at the strategic level to procedures for sharing data at the operational and tactical levels and now, in this section, to hardware and software needs (for example, network security protocols) at the technological level.

Currently, the United States uses the Contingency Theater Air Planning System (CTAPS) as its force-level planning system. The United States is also developing the Theater Battle Management Core System (TBMCS) as its follow-on force-level planning system. The NATO Consultation, Command, and Control Agency (NC3A) has developed the Interim CAOC Capability (ICC) software for force-level planning, and NATO has recently awarded a contract to a consortium led by Thomson-CSF for a follow-on system known as the Air Command and Control System (ACCS).

Although much interest has been expressed in defining interoperability requirements and standards between these systems, a number of factors complicate this process, including differing objectives, practices, and procedures; an increasing need for dynamic planning; the development of a COP; and the use of standards.

## Differing Objectives, Practices, and Procedures

**CTAPS.** The concept and design of each of these systems have proceeded with different objectives, practices, and procedures in mind. CTAPS has been the U.S. operational system since shortly after Operation Desert Storm. It is meant to accommodate large air campaigns of up to 3000 sorties per day. The design of CTAPS (and its companion intelligence module, the Combat Information System [CIS]), is based on a large number of interconnected modules, each devoted to a particular function within the air campaign planning process (preparing target folders, planning tanker orbits, etc.). CTAPS is also designed for a relatively large (600 to 800) AOC staff, where each staff member might be responsible for operating a small number of functional modules. Each module has a private database, and limited data are passed from one module to the other as common data. As a result of these factors and the integration of intelligence data within the process, the functionality and computational power of each module can be made relatively complex. At the same time, however, training and ease of use are significant challenges to the implementation of CTAPS with inexperienced personnel during wartime.

**TBMCS.** TBMCS is aimed at expanding the functionality of CTAPS by adding more integrated intelligence data and computational models to the ATO planning process. TBMCS also has consolidated the number of databases to two: the Air Operations Database (AODB) and the Modernized Integrated Database (MIDB). This move toward more common databases should enable TBMCS (in principle) to allow near-real-time updates to common data by a number of modules so that different modules have access to data earlier than if the process were serial, as in CTAPS. Accompanying this database consolidation is a replication engine designed to copy the AODB and MIDB regularly between different sites (e.g., AOC, air support operations center [ASOC], wing). Mission reports (MISREPs) are filed by the wings via a messaging system that updates the databases at the AOC. Much of the design of TBMCS is based on a relatively free exchange of information between different portions of the planning process, with the exception of intelligence production. Here, a suite of applications is responsible for the development of top-secret information (such as imagery) on a secure network ring, and secret-level data are passed down via National Security Agency–approved security guard software (e.g., Radiant Mercury for text and ISSE Guard for imagery).

**ICC.** The current NATO system, ICC, is substantially different from U.S. systems in its doctrine, practices, and procedures. In NATO, the CAOC is not quite analogous to the AOC in that the CAOC is primarily a battle management organization (as opposed to a C2 organization). As a result, the responsibilities for many force-level planning functions that rest with the U.S. concept of an AOC actually rest with higher-level headquarters in NATO. These differences are borne out in the design and implementation of ICC.

First of all, ICC is much more lightweight than CTAPS or TBMCS. It is designed for between 200 and 1000 sorties per day and includes little support for the integration of intelligence information. It is organizationally oriented and is designed principally for functions such as the allocation of sorties and the deconfliction of airspace. ICC is designed for a relatively small CAOC staff, who may be generalists with responsibility for a large number of functions within the planning process; hence, the software is aimed at "ease of use" as opposed to "depth of function." ICC includes little capability for the preparation of target folders, the integration of threat data, or weaponeering. In part, this reflects NATO practices and procedures, where most intelligence sharing is negotiated at higher echelons of command or where weaponeering calculations are made within national wings.

**ACCS.** Basic specifications for the forthcoming NATO ACCS have been available for some time and include the integration of ground radars and sensor fusion posts, which makes ACCS similar to the U.S. Theater Air Control System (TACS) in scope. However, the ACCS force-level module specification calls for a primarily air defense–oriented functionality with limited offensive planning capability. ACCS is planned to be linked with the NATO Battlefield Information Collection and Exploitation Systems (BICES), an evolving intelligence-sharing system for NATO designed to bring together information from a number of existing systems. The extent to which ACCS specifications may change, given the desire that NATO participate in more out-of-area operations, remains unclear.

**Interoperability Issues.** Recognizing that two force-level systems exist for the United States and for NATO, each of which accommodates different practices and procedures, is a crucial step in selecting an appropriate way to harmonize the programs to achieve interoperability. Currently, interoperability issues between CTAPS and ICC are being addressed by a wide variety of players, including the NATO Air Command and Control Management Agency (NACMA), NC3A, the USAF Electronic Systems Center (ESC), the U.S. J6, and the United States Air Forces in Europe (USAFE). To date, the solution has been to define messaging standards for communicating between systems at a basic level. Most significantly, standard ATO and ACO formats have been advanced (ATO-98 format) to allow for the exchange of ATOs (and some other messages, such as MISREPs) between the two systems. Although the necessary changes have yet to be fully implemented and fielded in both systems, negotiations will need to continue to ensure that subsequent changes to data formats by the United States or NATO are carried through to the messaging standard. Given the capabilities of both systems at present, messaging interoperability and the exchange of ATOs are likely to constitute an appropriate level of interoperability.

For ACCS and TBMCS to be used in support of contingencies ranging from small peacekeeping operations to MTWs, they must be scaleable. A systems architecture for software that enables the use of different modules by different users in different contingencies might allow significant flexibility in the functionality or capacity of a force-level planning system. Successful development of such an architecture requires major investments in modular software engineering and configuration management techniques. In some cases, simple workarounds (such as loading ATO parsing information on a laptop computer and linking to basic office automation tools) can be somewhat effective. Only modest progress in these areas of computer science have been made. Robust and flexible software systems of the future should incorporate these advances in order to minimize DoD's sunk software development costs and maximize the use of these systems by the warfighter in a range of contingencies.

**Cost Implications**. The cost of these force-level planning systems is relatively low. The Air Force has spent approximately $350 million on the TBMCS program to date, with another $100 million budgeted over the next six years. At the same time, the United States is funding approximately $170 million of the $700 million cost of the ACCS program. This commitment has been made over several years, so the annual costs are in the few tens of millions of dollars.

U.S. participation in the ACCS program provides for U.S. insight and influence on the direction of the program. This U.S. contribution can be viewed as an expenditure to ensure interoperability of U.S. forces with those of NATO in future operations. In the absence of compatible systems, laborious and often inefficient workarounds have to be devised, often at some cost in force effectiveness.

Given the importance of force-level planning and monitoring systems, the United States should leverage its investments in ACCS and TBMCS to ensure that they are interoperable. An agreement on a common ATO format is an important step in this direction.

## Dynamic Planning

In the future, however, as the capabilities of both systems expand to include the possibility of dynamic ATO planning beyond deliberate planning, the level of interoperability required may increase substantially. In this case, increased requirements for timely data will most likely require a number of capabilities. First, communication channels between force-level planners working from different locations or at different levels of classification will be required. Multilevel security network devices will likely be needed to facilitate these communications. A number of candidate architectures (such as double-encryption or virtual private networks) and data filters (Radiant Mercury, etc.) exist for these requirements, but both the United States and NATO must be convinced that the benefits of automated data sharing outweigh the operational risks of potential security compromises.

Clearly, these policy decisions must take into account more important strategic considerations in addition to the technological ones described here. Linking the U.S. MIDB with intelligence-sharing data systems such as NATO BICES—which comprises data from Linked Operational Intelligence Centers Europe (LOCE), the Maritime Command and Control Information System (MCCIS), and the Crisis Response Operation in NATO Open System (CRONOS)—is another optional part of this process. Again, secure technical architectures exist for this purpose, but limitations developed at the policy level on comingling of U.S. and NATO data tend to dominate these considerations.

## Common Operational Picture

Integrating situational awareness data via the COP is another essential aspect of information sharing for dynamic air control. Currently this information is shared at the CAOC level through use of a single U.S. system (the Global Command and Control System [GCCS] COP) with large projection displays. But other users at lower echelons of command (e.g., wings) often receive these data too late to be useful for ongoing operations. In recognition of the fact that a variety of users may require near-real-time information on at least a subset of this information, an effort should be made to facilitate the exchange of situational awareness data (air, maritime, and ground) on a near-real-time basis.

Finally, the exchange of tactical data between the CAOC, airborne control and surveillance assets, and fighters is necessary and will become more so in coming years. Interoperable data and voice links such as Link 16 and Have Quick (or SATURN) are essential to completing the loop between these parties (see Chapter Nine).

## Standards

No discussion of system interoperability and information exchange is complete without a discussion of standards. Although standards discussions do, in fact, permeate many working groups on interoperability in the military and commercial sectors, it is important to recognize their value. Standards can add value where they simply codify an existing (or a negotiated) consensus on an operational issue that involves technology. Predicting the success or failure of standards is similar to predicting the relative success of military operations in that strategic and operational consensus among major stakeholders is required. However, standard setting is often difficult to achieve when it also involves economic interests.

Widespread, cost-effective standards rarely take the form of bureaucratically produced documents that later become products made by several parties. Instead, the most popular standards are tantamount to specific technologies with well-identified "owners" who have an economic interest in being responsive to their user base and in changing the standard or technology accordingly. This "economic interest" or profit motive is important when a large user base is to be addressed.

In some cases, such as HTML and LINUX, professional recognition and fame (not necessarily economic gain) can be important motivators for technology creators, but usually only in the early stages of development. This "pride of ownership" is vital in the early stages of technology, where fluidity in the overall direction of the technology may be required. In fact, the temptation for overregulation of technology development at an early stage tends to stunt the success of the product in many cases. Once a technology does gain clear support (through a de facto or informal standardization process), significant resources are usually required to bring it into the mainstream.

Simply stated, standards are most effective when they codify existing consensus, or represent de facto public consensus, such as in the case of a popular product. By contrast they tend to fall flat when they serve as a surrogate for the consensus-making process or when the interests of key stakeholders are ignored.

## OBSERVATIONS AND SUGGESTED ACTIONS

Although NATO has modified its strategic focus to include non–Article 5 operations, CJTF doctrine and concepts of operations are still evolving. Current constructs represent potential sources of divergence from established U.S. practices and procedures. For example, they may not adhere to the C2 principle of centralized control and decentralized execution that governs U.S. air operations.

The Air Force should thus help NATO develop the CJTF concept of operations (CONOPS), associated processes, expert personnel, systems, and information-sharing protocols for out-of-area operations. In particular, the Air Force should ensure that the key doctrinal concept of centralized control and decentralized execution, which is inherent in U.S. joint-service air CONOPS, is institutionalized in the NATO CJTF concept.

Further, leveraging its expertise and capabilities in planning and executing air and space operations in power projection missions, the Air Force should man key positions in the emerging deployable and key static CAOCs to maintain the influence of U.S. practices and procedures in NATO. It should also develop and maintain a cadre of experts who can provide support to higher NATO headquarters (if needed) to help develop air campaign plans and assist in execution monitoring.

Information sharing is also a major interoperability challenge. This includes establishing intelligence-sharing policies at the strategic level, defining procedures for sharing operational and tactical data, and developing hardware and software capabilities at the technological level to facilitate information sharing among systems. In many cases, strategic policy considerations negate lower-level solutions and workarounds. For example, the sharing of tactical data (e.g., targeting data, ATO information, and ACO information) that should normally occur at the tactical level may become a matter for negotiation at higher echelons.

The Air Force should help NATO define the desired level of information sharing between planned U.S. and NATO force-level planning and execution-monitoring capabilities (organizations, procedures, personnel, and systems). This can be accomplished during the development of the CJTF CONOPS, which should clearly define the information exchange requirements for scenarios involving the NATO CJTF and U.S. coalitions.

Because of the importance of force-level planning systems, particularly to air operations, interoperability between U.S. and NATO systems is crucial. Given the current capabilities of TBMCS and ACCS, messaging interoperability and the exchange of ATOs are likely to constitute an appropriate level of interoperability. As these systems evolve to support advanced warfighting concepts such as dynamic planning and execution, the level of interoperability will likely increase substantially.

Therefore the Air Force should first define the desired level of interoperability between TBMCS and ACCS and between TBMCS and ICC, and it should then ensure that this level has been achieved. At a minimum, a set of common messaging standards for information exchange should be defined for TBMCS and ACCS. The Air Force should also develop incentives that facilitate dialogue between the TBMCS system program office (SPO) and the ACCS program director (NACMA) or directly between contractors (Lockheed Martin and Thomson-CSF/Raytheon).

Chapter Six

# SPACE DEVELOPMENTS

Historically, the predominance of U.S. investment in and experience with space systems has minimized the consideration of space as an area with real or potential interoperability problems. Whether with select allies or within a broader Alliance framework, the United States has provided the bulk of products or services derived from space assets, especially satellite reconnaissance data to support military coalition operations. The development of the Global Positioning System (GPS) has also allowed the United States to dominate space-based positioning, navigation, and timing.

Yet while U.S. sharing of space-based data has increased over time, some European allies are dissatisfied with the nature of the information shared (in terms of volume and levels of analysis) as well as with the lack of European input into tasking mechanisms. Moreover, the Europeans have ever greater expectations about the value of space, whether for civil resource management or defense purposes.

These factors and the proliferation of space and space-related technologies have created new incentives for the development of new space capabilities, especially within Europe. While the United States will continue to dominate space for some years to come, European space developments may provide important opportunities to improve the security advantage of the United States and/or its allies.

Space may be a new area in which to consider interoperability, either as a source of progress or as a source of potential conflict. For example, if different sources of space-derived data are used by the United States and its European allies, very different viewpoints can arise when each independently performs critical military functions such as indications and warning, target development, and battle damage assessment.

The United States needs to seriously consider the existing and planned slate of European space capabilities—and their associated ground segments—for their potential contribution to coalition operations, taking into account interoperability considerations. This involves consideration not only of the interests and capabilities of individual nations but also of developments within European space cooperation. And space, as it increasingly becomes valuable as a source of diverse kinds of information, must be considered in the broader context of information sharing, not only among U.S. government agencies but also with allies in a range of military operations.

Thus, the case study described in this chapter highlights areas of potential cooperation or competition that might arise between U.S. space programs that have proved vital in past coalition operations, and Europe's growing desire to increase its nascent space capabilities as a means of lessening its dependence on U.S. assets in future military operations.

## THE BROADER CONTEXT FOR SPACE COOPERATION

Space cooperation should be considered in the context of overall U.S. national objectives as well as the increasing information value that space provides and the forces that accompany the larger set of information-sharing issues and options.

U.S.-European military space activities suffer a plight equal to if not worse than the slate of more traditional interoperability issues that arise among the NATO allies. As we will see in our discussion of the various European space programs, the United States and Europe are divergent in overall strategy and perspective as well as in the nature of public-private interactions with regard to technologies and markets. Both are important as they relate to the use of space in military affairs, to the scope and speed of innovation in space technologies, and to the information value derived from such technologies on the ground.[1]

Because the value of space is so closely wedded to its provision of information, we must also consider space cooperation within the broader context of U.S. information sharing. One aspect of U.S. strategy that has not been sufficiently explored is the extent of U.S. information sharing that exists in both classified and unclassified domains, as well as the overall cost-benefit analysis that is derived from that sharing or any inconsistencies in those sharing arrangements.[2]

Information-sharing policy assessments need to be informed by an objective appreciation of what kinds of space-based or other data military organizations process and how. This is often a function more of organization, information culture, and experience with specific data sets than with information policy per se. For example, one can compare the substantial U.S. experience in all-source integration of information sources, including imagery data, in the U.S. defense program with Europe's fledgling efforts to fold information derived from satellite images into European military planning and operations.

The rapid emergence of commercial space and information markets also promises to advance lagging countries and entities to the extent that they are willing to rely on commercial sources within their military programs.

## MOTIVES AND METHODS FOR SPACE COOPERATION

U.S.-European space cooperation is not without precedent. European countries—especially France and the United Kingdom—rank prominently on the list of countries with which the United States has international agreements in remote sensing for civil and scientific purposes.[3] Meteorology, wildlife tracking, climate change,

---

[1]See Gompert et al. (1999), p. 9.

[2]Such an analysis is beyond the scope of this effort.

[3]See Wagner (1998).

atmospheric science, ocean science, land use, and mapping are among the themes covered in these agreements.

There are a number of additional motives for the United States to consider cooperation with its European allies in space:

- U.S. National Space Policy (PDD-49) encourages cooperation in space-related activities for achievement of scientific, foreign policy, economic, or national security benefits for the nation.
- The U.S. Policy on Foreign Access to Remote Sensing Space Systems (PDD-23) and its related Commerce Department regulations recognize increased foreign interest in space systems as well as the opportunity to enhance U.S. industrial competitiveness and security through foreign access to remote sensing data and technology.
- U.S. Space Command, in support of the warfighter, envisions growth in partnerships with foreign militaries.
- The interests of DoD and the intelligence community dictate cooperation in space and space-related areas, such as (1) promotion of bilateral ties and Alliance cohesion, (2) promotion of doctrine, operations, and equipment interoperability, and (3) information and intelligence sharing.

Space cooperation, according to the new DoD Space Policy, is to be undertaken when mutual, tangible benefits are available in support of the strategic enabling function of space.[4] Given its own investments, its expertise in space operations, and its use of space-derived information, the U.S. Air Force—in collaboration with other DoD and U.S. government agencies—should help shape space cooperation with the U.S. European allies.

There are other dimensions, such as foreign motives, to consider if cooperation is to be successful:

- Foreign governments have a variety of interests in space, including foreign policy and national security, international prestige, autonomy in national decisionmaking, creation or maintenance

[4]See DoD (1999b).

of their industrial base, land use and resource management, and political motives.

- While foreign governments want access to U.S. information and space technology, they are also trying to create indigenous capabilities.
- Governments are looking increasingly to strong growth in commercial space capabilities (in areas like communications, remote sensing, and navigation) as an important component of their use of space.[5]
- The development of foreign space capabilities is seen as offsetting possible U.S. and other international political agendas, as they provide an independent view of world events. Of course, unless they choose to take advantage of the commercial space market, foreign interests in space generally remain bounded by the high expense of developing, launching, and operating space systems.

Figure 6.1 depicts a conceptual spectrum of the different ways that governments might choose to cooperate in space programs. Cooperation can range from the sharing of data at the lower end of the spectrum to coproduction and codevelopment of space systems and their associated ground infrastructure at the higher end of the spectrum. Each of these methods carries political, technical, and security benefits and risks.

There are limits to space cooperation. Shifting U.S. views on technology transfer, in particular satellite technology transfer, will limit cooperation in this area. Space cooperation also requires some agreement on the operational decisionmaking structures for satellite systems, especially as they relate to access and control of data (e.g., prioritization of imagery tasking). Data policy and intellectual property-rights issues are at stake, yet such issues either are unresolved in international forums or are a potential source of

---

[5]For specific details on the state of global commercial space activities, see Space Publications (1999).

RAND MR1235-AF-6.1

| Space System Activities | Examples | General Features |
|---|---|---|
| Products | • Data sharing—governments trade or share intelligence products<br>– Imagery for analysis trades, etc. | Coupling Trend ↓ |
| Operations | • Joint operations—governments jointly operate intelligence assets<br>– Bilateral/multilateral system use<br>– Technical support | |
| Acquisition | • Procurement—governments purchase systems developed and produced by aerospace contractors in partner countries<br>– Systems sales<br>– Systems leases<br>– Contingent sales<br>• Production—aerospace contractors from partner countries produce systems developed in one of the partner countries<br>– Coproduction<br>– Licensed production<br>• Development—aerospace contractors from partner countries jointly develop and produce space systems<br>– Codevelopment<br>– Technical support | |

**Figure 6.1—Space Cooperation Methods**

conflict between the United States and its partners.[6] Finally, there are no clear valuation metrics for cooperation.

European space policies and programs have some unique challenges. Space issues in Europe are dominated by industrial considerations, although some of this has abated as European aerospace consolidation has taken place. Fidelity to European policies of *juste retour*—an expectation that Europeans would receive industrial work shares consistent with their governments' investment in European cooperative programs—is difficult to satisfy. Individual countries will have their own preferences and capabilities for cooperation, which must be considered against the slate of overall U.S.-European cooperation in this area. As is the case with the United States, government-to-government cooperation will have an impact on the emerging market for commercial space products and services, which

[6]See Pace et al. (1999).

may affect the market in disadvantageous ways for U.S. or European industry. Finally, European governments may be looking increasingly to space systems as a source of cost-offsetting revenue—this is part of the current rationale for Galileo—that will conflict with U.S. data policies and may disadvantage U.S. interests.

## EUROPEAN SPACE PROGRAMS

We identified five European space programs that highlight the choices for the United States on whether to cooperate or compete with Europe in space technologies and operations. These programs also highlight the diverse opportunities and challenges involved in maintaining the interoperability of U.S. and NATO allies' space-derived services and products. Our research endeavors to answer the following questions concerning each program:

- **French space program.** How should the United States respond to continuing French interest in greater military space cooperation, particularly given France's unwillingness to appear dependent on the United States in key military technology areas?
- **Italian space program.** Italy is pursuing a highly ambitious dual-use remote sensing satellite program in a bid to increase its importance among Europe's advanced technology leaders. Should the United States encourage this development even though SKYMED/COSMO (Constellation of [Small] Satellites for Mediterranean [Basin] Observation) program, while offering the opportunity for burden sharing in military operations, creates added competition for American commercial remote sensing firms?
- **WEU Satellite Centre.** The WEU Satellite Centre provides a nascent image processing capability to support European civil and security-related activities. Can the United States and its NATO allies leverage WEU Satellite Centre capabilities in coalition operations?
- **Galileo program.** The European Union has strongly endorsed plans for the Galileo satellite navigation system as an independent source of global positioning data. What might be the interoperability implications of this alternative European architecture for the United States and its NATO allies in coalition operations?

- **European space-based radar programs.** There is a strong European interest in new imaging phenomenology, particularly those associated with radar and hyperspectral imagery. This interest is reinforced by the Allied Force experience. Would interoperability in coalitions be enhanced by more U.S. cooperation with Europe on radar and hyperspectral technologies and applications?

We discuss each of these programs in more detail.

## French Space Program

France and the French Space Agency (Centre National d'Études Spatiales, or CNES) dominate European space programs. France has the largest and broadest space program in Europe, largely in areas of earth observation, communications, microgravity, and satellite navigation.[7] France also serves as the greatest proponent of European cooperation in space, in part as a response to U.S. space activities and consolidation of the U.S. space industrial base. The French have supported European cooperation both in the civil arena—in the European Space Agency (ESA)—and in the defense arena, as in satellite communications and remote sensing.

Unlike the U.S. model for organizing space activities—general separation of civil space and military space—France and most other European nations manage their space programs under one organization such as CNES.[8] In recent years CNES has managed programs related to both civil and defense missions and has undertaken a consolidation of its own programs. One of these, the Helios program, is representative of the challenges of space cooperation as well as the possibility of conflict and cooperation as they relate to U.S.-European strategic purpose and interoperability.

Helios is the first in a family of military observation satellites to provide a limited daytime surveillance capability.[9] It was launched

[7]See CNES (1999).

[8]The comparatively smaller size and scope of European space programs facilitates this.

[9]See McLean and Swankie (1998).

from French Guiana in July of 1995 with a life expectancy of four or five years. Helios was designed to fulfill a long-standing desire on the part of the French government for space-based imagery and was also intended to quell that government's historic frustration over Europe's reliance on the United States for such data in the face of unsuccessful attempts at space cooperation dating back to the 1970s.[10] French interest in space-based military remote sensing was further stimulated by their experiences in the Gulf War.

In addition to satisfying its own needs from Helios, Paris has attempted to make the Helios program representative of the opportunities for further European space cooperation, with mixed results. Political and financial considerations prompted the French to solicit broader European support for the program in the late 1980s, resulting in Italian and Spanish commitments to finance approximately one-fifth of total system cost.[11] However, both Italy and Spain are considering alternatives to France's follow-on program, Helios 2, because they have doubts about whether that program is the best way to meet their future defense requirements.[12]

A subsequent initiative captured German involvement in the Helios program and responsibility for the development of a related radar program known as Horus, which ultimately failed as a cooperative program largely because of German fiscal constraints.[13] Yet France and other Europeans cite Helios' primary value as a capability that is independent of the United States. And, as the Europeans focus anew on their common foreign and defense policy needs, remote sensing capabilities are seen as vital to European intelligence needs and overall strategy. One strategy may be to continue to build remote

---

[10]Ibid.

[11]Italy's contribution is 14 percent, and Spain's contribution is 7 percent (de Selding, 1995a, and de Selding, 1999).

[12]See de Selding (1999).

[13]The fiscal constraints are discussed, for example, in de Selding (1995b) and de Briganti (1995). After this research was completed, France and Germany decided to resurrect plans for a joint military reconnaissance satellite program based on the French Helios and a new, lower-cost (approximately $340 million) German radar imaging program known as SAR-Lupe (CNES, 2000; de Selding, 2000). Italy may also join this new pact, which could provide Europe with a robust and independent space-based reconnaissance capability (Lewis, 2000).

sensing capability as a means to stimulate U.S. interest in greater cooperation.

With regard to the United States, the French are interested in reestablishing and reenergizing an early 1990s space cooperation agreement, with an emphasis on satellite communications, navigation, and remote sensing. French discussions have centered on the need for greater synergy among U.S. and coalition space assets, given the increasing role of coalition operations. French officials have also indicated a desire to remain linked to U.S. plans, given the relatively large scale of those programs and the long time lines associated with the development and production of French space systems.

French views on space cooperation with the United States include the concept of balance between sovereign and cooperative needs—a classic French concern—as well as focus on cooperation in areas throughout the space infrastructure (in other words, space and ground segments). French officials see little room for cooperation in areas that have been commercialized, such as launch. They also envision themselves as a potential intermediary between the United States and a broader set of European space partners and activities.[14]

The French desire for cooperation with both the United States and the Europeans is consistent with other defense activities: as much as 30 percent of the French defense budget is devoted to cooperative programs.[15] Yet French pursuit of space cooperation with the United States has strong political motives as well: France looks to space cooperation with the United States not only as a measure of its prominence but also as a means of offsetting the challenges the Helios program faces and as a way to remain linked to U.S. space plans and technologies. At the same time, France will continue to pursue pan-European space initiatives.

With regard to future interoperability, one implication of U.S.-French space cooperation is that it will increase the number of sources of data and information—including not only Helios imagery

---

[14]This discussion is based on a number of direct interactions that one of the authors has had with French Embassy officials in Washington, as well as representatives of the Délégation Général pour l'Armament (DGA) in France and the United States.

[15]See International Institute for Strategic Studies (1998).

but also commercial assets such as Satellite pour l'observation de la Terre (SPOT). The U.S. military's experience with access to and use of SPOT data has been positive and is expanding. Unless data sharing is properly synchronized, however, it can lead to conflicts in indications and warning; target nomination, approval, and development; and battle damage assessment.

France has occasionally cited Helios data as a source of information that differed from the stated public U.S. view of some situation.[16] France has also suggested that U.S. sharing of information is largely tailored to U.S. political aims rather than to a comprehensive view of a particular situation. This issue was raised during the Kosovo conflict, in which the French expressed concern with the substance and volume of U.S. assessments. Competing data sources require some venue for discussion of alternate interpretation, perhaps below the political level. Besides reducing the probability of conflicts, properly synchronized collection may also reduce the burden on U.S. collection in future Alliance or coalition operations.

At a minimum, in collaboration with DoD and other government agencies, the U.S. Air Force should closely monitor French space developments, as well as the progress of the U.S.-French space cooperation initiative, and then clearly define information-sharing protocols for those space assets that support future coalition operations.

## Italian Space Program

Italy has initiated the development of an ambitious remote sensing satellite program known as SKYMED/COSMO. Although one of the partners behind the French-led Helios military reconnaissance satellite program, Italy has nevertheless decided to pursue its own remote sensing program in an effort to become a leader in European remote sensing capabilities.[17] Italy sees an important strategic role for itself

---

[16]France argued in December 1996 that the U.S. interpretation of developments in Iraq were different from French interpretations of Helios imagery, thereby diminishing a U.S. rationale for military strikes at that time.

[17]See Taverna (1999).

and within NATO in maintaining awareness of developments in the Mediterranean and North Africa.

The SKYMED/COSMO plan calls for a constellation of seven smaller imaging satellites (600 kg each), with three optical satellites and four radar imaging satellites. The radar satellites would feature a synthetic aperture radar (SAR)-2000 sensor capable of about 1-meter resolution, while the optical sensor would have a 2.5-meter resolution.

During the past two years, Italy has been exploring options for industry-to-industry cooperation, including cooperation with U.S. aerospace partners. If a government-to-government agreement can be reached, this approach will enable Italy to acquire more advanced satellite remote sensing capabilities than might otherwise be available to it on the commercial market. Like France, Italy is seeking European partners for the SKYMED/COSMO program to help spread the estimated $550–$600 million acquisition costs and reduce the program's technology risks.

Italy appears determined to enhance its standing in the space field, both within Europe and on an international scale. Its estimated government spending on space for 1999 is over $700 million, which ranks Italy as fifth in the world in nonmilitary spending on space. Its space budget is comparable to Germany's and is substantially more than Russian spending on nonmilitary space. Italy is also raising its international prominence by directly contributing to the International Space Station program with national resources beyond those committed through ESA.

Italy's ambitious plans for developing its own space capabilities create several opportunities for cooperation with the United States that could yield significant benefits for both sides, particularly in terms of providing imagery data of mutual interest on locations within the politically volatile Mediterranean region.

One potential area for U.S.-Italian cooperation is the Discoverer II satellite program. Italy could contribute additional resources and technical expertise in developing smaller radar imaging satellites. As various aspects of Discoverer II are under consideration, more thought needs to be given to how international collaboration might affect cost, technical, and data aspects of the system.

NATO combat operations against Yugoslavia highlighted many of Italy's major security concerns. These operations indicated the broader trade space within which space cooperation must be considered. Space access, in essence, may be valuable to a potential partner in return for other kinds of access (landing and basing rights, other kinds of intelligence data). In addition, Operation Allied Force and continued surveillance of the Balkans provide opportunities for greater U.S.-Italian cooperation. Hence, the U.S. Air Force should take advantage of existing coalition information-collection operations in the Balkans to gain a more focused appreciation of the leeway of Italian government agencies and aerospace firms for cooperating with the United States regarding military space.

## WEU Satellite Centre

The WEU Satellite Centre was created by the WEU Council of Ministers in 1991 and declared a "permanent WEU body" in 1995. Located in Torrejón, Spain, the Centre was declared fully operational in May 1997. The satellite center is an indigenous image-processing center that relies on commercial and European military sources of data to support the civil and military information needs of WEU and associated members. The Centre's self-described mission is "to exploit imagery derived from space observation satellites for security and defense purposes."[18]

Among the data sources exploited by the Centre are the following: (1) primarily commercial imagery sources (e.g., SPOT, Earth Resources Satellite [ERS-1]), (2) Helios imagery contributed by France, and (3) occasionally, airborne imagery provided by member nations. Imagery is acquired in response to questions posed by WEU member nations and any others approved by the WEU Council. Figure 6.2 portrays the work process for imagery collection, interpretation, and dissemination for the Centre.

The products developed in response to these WEU taskings are provided in the form of a hardcopy "dossier" that contains the imagery

---

[18]See WEU Satellite Centre (1998).

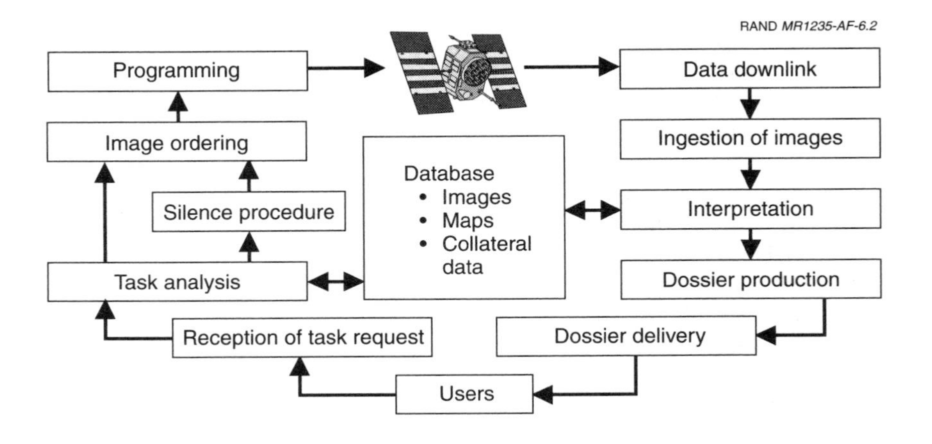


**Figure 6.2—WEU Satellite Centre Imagery Cycle**

and maps used in the interpretation as well as the analytical markings and interpretation performed at the Centre.[19]

The Centre's processes are interesting from a number of perspectives. First, the level of image interpretation done at the Centre is fairly rudimentary (similar to U.S. phase one exploitation: identification of specific structures and activities); more advanced phases of exploitation are considered political judgments and are done largely in the national capitals. Second, because of the multinational character of the Centre, analysts are drawn from different WEU countries and are therefore somewhat representative of the views of those countries. In essence, the Centre is an interesting laboratory in which analytical interpretation and political judgment intersect. Third, the dissemination policy of the Centre is that its products are available to all WEU and associate members, so there are no problems with compartmentalization and selective sharing under this construct.

Recent events in Kosovo have catalyzed WEU interest in expanding activities related to space-based reconnaissance, and the Satellite Centre is firmly engaged in those activities. While earlier studies by

---

[19]Ibid.

the WEU on the feasibility of acquiring a shared European space reconnaissance system concluded that the costs were too high,[20] Operation Allied Force has reenergized political interest in such capabilities.

Various studies are under way within the WEU to look at options for increasing access to space observation, improving the number and quality of European image analysts, and creating electronic access to the Centre's product.[21] These studies indicate a desire to leverage existing and planned European national programs such as Helios and the civil programs sponsored by ESA. Independent of these activities, remote sensing commercialization—including the successful operations of Space Imaging's Ikonos satellite—means that the Centre will have access to increasing amounts of quality imagery data.

The WEU Satellite Centre offers the potential for burden sharing in military operations. Whether as a source of additional imagery data or analysis,[22] the Centre could support some U.S. and especially European military operations. But while the Centre is gaining expertise in exploiting commercial imagery, the scope and depth of its knowledge are limited, and it has only a limited capability to support coalition military operations. The direct downlink and digital dissemination are among the current upgrade initiatives that would enhance the Centre's ability to support U.S. and allied military operations.

Image processing remains a challenge because of a lack of indigenous sources and varying national data processes and policies. Cooperation and commercialization will drive countries to common standards (e.g., Earth Resources Data Analysis System [ERDAS] image processing systems); some dedicated military will remain (OCAPI and PEPITE image processing systems). Meanwhile, the United States clearly has an overwhelming advantage in imagery exploitation and data use. The United States needs to know more

---

[20]See de Selding (1995c).

[21]Dissemination of product heretofore has been accomplished via courier or traditional postal means. See WEU Technological and Aerospace Committee (1999).

[22]A parallel to "imagery for analysis trades" could be envisioned and has an interesting precedent in the signals intelligence arena .

about current WEU procedures and future enhancements to determine the best approach for burden sharing and to develop an acceptable level of interoperability. U.S. Air Force experts on imagery analysis and targeting could help synchronize procedures and establish acceptable quality control procedures.

## Galileo Program

Space-based positioning and navigation are issues of growing importance to Europe for a variety of reasons related to civil requirements, industrial competitiveness, and security and sovereignty concerns about having to depend on the United States and other countries for this critical information source.

In a major step toward acquiring an independent satellite navigation capability, the European Union (EU) approved a study plan in May 1999 to begin the definition phase of the Galileo program in coordination with ESA.[23] This unprecedented display of EU support for a European space program was also a case of unprecedented coordination on a space program from Brussels as opposed to the national capitals. The strength of political support for Galileo was reflected in its oversubscription by voluntary contributions from ESA members.

The Galileo program, a major technology and infrastructure program, involves extensive cooperation between ESA, which is responsible for the space segment, and the European Commissioners for Industry, Research, and Transport, which will manage the extensive ground infrastructure. At an estimated total cost of $2.5–$3 billion, the Galileo program is expected to involve some 21 to 36 satellites in medium earth orbit (MEO) and an additional 3 to 9 satellites in geostationary orbit (GEO). Although envisioned as a full positioning system that is independent of GPS and Russia's GLONASS, Galileo still emphasizes the need for interoperability with existing systems.

The Galileo navigation satellite system, also known as Global Navigation Satellite System 2 (GNSS 2), builds on the European Geostationary Overlay Service (EGNOS). EGNOS groups together the first generation of European positioning satellites as the European

---

[23]See Barensky (1999) for an overview of European motives for Galileo.

regional contribution to GNSS 1 provided by the U.S. GPS and the Russian GLONASS system. The EGNOS system, which consists of additional satellites and ground-based systems to become fully operational by 2002, is designed to improve navigational accuracy to 5–10 meters,[24] to monitor the integrity of GPS signals, and to ensure the line-of-sight availability of at least six satellites at all times.

The strong political endorsement for proceeding with the Galileo plan indicates the EU's determination to avoid depending on other countries, such as the United States and Russia, for critical data to meet its positioning, navigation, and timing (PNT) requirements. These PNT services are currently free, but European planners view the Galileo program as important to ensuring that Europe is not vulnerable to future changes in the availability of service or the imposition of fees. These planners also see their own PNT capability as critical to economic development and to improving European industry competitiveness on the international market. Satellite navigation is viewed as important in safe and efficient navigation, particularly for air traffic control, and in many economic enterprises. Without something like the Galileo program, Europeans fear they will fall behind the United States—and to some degree Japan—in taking full advantage of this important information technology.

Whether or not the Galileo program is ever fully realized, this effort will likely stimulate major disputes with the United States and others in the trade and military realms in spite of European pledges to seek compatibility with GPS and GLONASS. A major source of conflict arises from the competition for spectrum that is caused by European interest in developing an independent satellite navigation and positioning system. Competition between GPS and the new mobile satellite systems (MSSs) for high-quality spectrum could result in degrading the quality and reliability of the GPS signal. This could lead to market fragmentation for civil frequencies.

---

[24]As of 2 May 2, 2000, the United States no longer degrades the GPS signals available to the public (White House, 2000). With Selective Availability turned off, the geopositioning accuracy for the Standard Positioning Service is now less than 5 meters (GPS Support Center, 2000). This decision occurred during the final preparation of this report, and thus we have not assessed its impact on European space-based positioning and navigation programs such as Galileo.

Resolving liability and other legal issues is complicated by the disparate public-private philosophies that both sides bring to this question. The U.S. approach views GPS signals as a public good that is made available without fees. By contrast, the European approach will impose charges for restricted access to higher-accuracy service and will emphasize private enterprise participation to reduce program costs.[25] There is also an important difference in how both sides view liability and what types of public warnings are necessary.

Europe's decisions on Galileo also have important implications for sustaining Russia's GLONASS system, which is steadily degrading over time without the needed replacement satellites and ground modernization. Despite its growing operational problems, however, the GLONASS system makes use of valuable spectrum allocation that has a strong appeal to Europe.

Given the U.S. Air Force's substantial ongoing investment in the GPS system and in GPS-guided weapons and delivery platforms, the issue of GPS integrity and availability is critical to future Air Force planning and operations, particularly in the European region. Although European leaders see Galileo as a potential revenue source or cost offset, it is not clear that full thought has been given either to European military implications or to those for U.S. and allied operations.

Ongoing U.S. negotiations with the Europeans over Galileo highlight the need for compatibility among the navigation systems as well as for longer-term discussion about how military systems may be affected by the existence of Galileo. The U.S. Air Force must maintain an active role in these discussions in order to help U.S. authorities understand, identify, and protect U.S. equities.

## European Space-Based Radar Programs

Europe is strongly committed to pursuing space-based imaging radar technologies for a broad range of applications. This emphasis on radar imaging satellites is reflected both in national programs and in the Earth observation programs that are supported by ESA. The chief

---

[25]See de Selding (1995c).

ESA remote sensing program focuses on the European Remote Sensing (ERS-1 and -2) satellites launched in 1991 and 1995. These satellites have microwave instruments, including a SAR, that are used for environmental monitoring, ocean research, and disaster monitoring. A much larger Earth observation satellite, Envisat-1, is scheduled for launch in 2000. It will have an advanced SAR and other sensors.

Several countries are also pursuing national programs on radar technologies. Among the most notable is Italy's SKYMED/COSMO program, described earlier. Germany has extensive experience with radar imaging technologies. Deutsche Aerospace was the prime contractor for the ERS satellites. Following its decision not to proceed with joint development of the Horus radar imaging satellite with France—mainly because of budgetary constraints—Germany initiated new radar imaging satellite efforts, including studies on a smaller series of radar satellites known as SAR Lupe and TerraSAR, a joint U.K.-German development to develop a radar satellite with a 1.5-meter resolution that could be launched in 2004.

German interest in radar imagery is driven by diverse requirements. Scientific research has included sophisticated experiments to derive SAR interferometric data from tandem flights of the ERS-1 and ERS-2 spacecraft. Given frequent cloud cover and adverse weather, radar imaging is also needed for various civil missions, such as disaster monitoring for flooding. Furthermore, SAR data are particularly useful for creating digital elevation models (DEMs) that can be used to create three-dimensional terrain visualization. Finally, German military involvement in military and peacekeeping operations, such as Kosovo and Bosnia, has created another imagery need.

The key point here is that Europeans have increasing stakes in radar and other advanced forms of remote sensing. The U.S. common experience in Kosovo reaffirms the need for radar capabilities. This, of course, is at a time when the United States is also pursuing enhanced space-based radar capabilities. The Discoverer II program and other radar programs are part of an overall improvement of ISR capabilities. Although the United States retains an overall advantage in advanced space capabilities of this sort, it must also interact increasingly[26] with new foreign-government and commercial initiatives that may be of value to U.S. or allied operations. Effective cooperation between the United States and Europeans is one way to offset the costs of the Discoverer II program, for example, and as well as potentially minimize interoperability issues in future military operations. Leveraging its expertise in radar imagery assets and as a partner in Discoverer II, the U.S. Air Force should be an active participant in U.S.-European space-based radar cooperation efforts.

## OBSERVATIONS AND SUGGESTED ACTIONS

The U.S. trade space with regard to space cooperation with Europe is multidimensional. U.S. options and strategies for space cooperation include bilateral and multilateral agreements, commercial and government activities, and cooperation along a spectrum from the sharing of space-derived data to joint development of space systems. The costs and benefits of any one of these arrangements should be weighed in light of its overall impact on U.S. objectives. This means that space cooperation dominated by U.S. contributions may provide the leverage to gain other non-space-related contributions from its NATO allies in coalition operations, i.e., access to airspace and infrastructure. NATO—a logical venue for some of these discussions—appears to be absent from most major discussions about space issues, especially within the broader context of interoperability.

Within this case study, we have attempted to discuss the broad dimensions of U.S.-European space cooperation with a view toward enhancing interoperability (or at least identifying areas of potential progress or conflict related to interoperability) in coalition operations. Aspirations of European space actors like France (the dominant player) or Italy and Germany (emerging players) could provide opportunities to improve interoperability, especially to the extent that U.S. and European decisionmakers develop standards and methods for cooperation in space-related capabilities to support

[26]Recent U.S. discussions with the Italians and Canadians reflect the kinds of challenges that the United States faces in this area. For example, the United States and Canada have been engaged in an occasionally contentious set of discussions over RADARSAT 2 (see Pearlstein, 1999, or Ferster, 1999).

military coalition operations. A good first step, especially in the case of France, would be to improve data-sharing protocols; this would precede any greater collaboration in space system development.

The WEU Satellite Centre represents a potentially important opportunity for information burden sharing. In exchange for increased access to WEU capabilities, the U.S. Air Force, in collaboration with other DoD and U.S. government organizations, could help the Centre train its imagery analysts and help the WEU develop quality control procedures that would optimize its products for military coalition operations. This collaboration would be an important step in developing a policy for NATO allies to use information developed at the Centre for coalition operations.

Although still in its definition phase, the Galileo program could negatively affect U.S. defense interests, given the possible need for additional technical capabilities to use the signal aboard U.S. military platforms as well as Brussels' interest in pursuing a fee-based service for using precision navigation signals. Galileo could also enter into direct competition with U.S. capabilities, although current negotiations reveal an understanding of the value of compatibility between the systems. Nonetheless, significant military policy and operational disputes could arise from competing political or commercial interests, as well as from the added costs of equipping or refitting military platforms with new navigation signal receivers capable of using signals from both the U.S. and European systems.

Finally, European political (especially post-Kosovo) and technical experience with radar means that there are potential opportunities for technology and other exchanges with the European space-based radar programs.

The United States will continue to derive comparative advantages from space, based on its strategic requirements and its decades-long investment and experience in using space for military and national security purposes. However, the five programs examined here clearly reflect increased European appreciation for the security and economic value of space as well as Europe's willingness to take independent action. Thus, U.S.-European space relations appear to be at a crossroads—with important ramifications for future cooperation or competition, and any attendant benefits or damages to interoperability and security relations.

Because the United States retains advantages of architecture in space, not to mention unparalleled investment and experience, the United States should be able to shape European partnerships (whether within or outside of NATO) in a way that benefits both U.S. and NATO allies' security interests. But because of the size of the trade space, this will require considerable thought about a strategy that invariably involves cooperation, competition, and some continued U.S. dominance. The U.S. Air Force needs to play an active role in such activities. In particular, it can contribute to the broader realm of policy and operations that relates to information-sharing practices—both products and services derived from space assets—for use in U.S. and NATO allies' coalition operations.

Chapter Seven

# AIR SURVEILLANCE AND CONTROL

Over the past 25 years, the U.S. Air Force has demonstrated the value of an airborne air surveillance and control system. As an airborne system, the E-3 AWACS can compensate for the limitations of ground-based air surveillance and control systems by reducing the effects of radar terrain masking, by extending low-altitude radar coverage, and by extending communications and control to forces operating beyond the range of ground control centers. As a mobile system, AWACS can compensate for the limitations of fixed surveillance radars and control centers by providing radar coverage, communications, and control where it is most needed, including over enemy territory, and when it is most needed, including early in a conflict before ground centers can be established.

Several U.S. allies have also recognized the value of such capabilities, as evidenced by their acquisition of their own E-3 fleets. Table 7.1 shows the number of AWACS owned by the United States, NATO, and U.S. European allies. The United States owns and operates 32 E-3B/Cs; France owns and operates four E-3Fs; and the United Kingdom owns and operates seven E-3Ds. The NATO Airborne Early Warning Force (NAEWF) consists of two components—the E-3A component with 17 aircraft owned and operated by NATO, and the E-3D component with six aircraft owned and operated by the United Kingdom (i.e., the U.K. has "declared" six of its seven E-3Ds to NATO; they are manned predominantly by U.K. personnel[1]).

[1]The U.K. has made an agreement with NATO that the six aircraft are available when NATO requests them if they are not otherwise needed by the U.K. for national reasons.

**Table 7.1**

**U.S. and European E-3 AWACS Fleets**

| Country | Number of Aircraft | Main Operating Bases |
|---|---|---|
| United States | 32 E-3B/Cs | Tinker AFB, OK; Kadena AB, Japan; Elmendorf AFB, AK |
| France | 4 E-3Fs | Avord, France |
| United Kingdom | 7 E-3Ds | RAF Waddington, England |
| NATO | 17 E-3As | Geilenkirchen, Germany |
| | 6 E-3Ds[a] | RAF Waddington, England |

[a]The U.K. has "declared" six of its seven E-3Ds to NATO.

The E-3A component is NATO's first and only fully integrated multinational operational unit, manned by personnel from 13 NATO nations.[2] The headquarters of the NATO Airborne Early Warning Force Command (NAEWFC) is collocated with the Supreme Headquarters Allied Powers Europe (SHAPE) in Mons, Belgium, and its commander reports to the two Major NATO Commanders (SACEUR and Supreme Allied Commander Atlantic [SACLANT]).

These AWACS programs demonstrate a long history of international cooperation among allied partners—both within the military and within industry—in developing and procuring nearly identical airborne systems for air surveillance and control. This commonality has allowed them to operate together in peacetime and in a number of military operations. For example, common systems and standardized procedures allowed NATO to combine aircraft of different nations in recent Balkan operations to accomplish military objectives. Thus, the AWACS programs provide one of the best examples of the

They have also been available for other coalition operations. For example, E-3Ds have been in Italy since 1993 supporting various operations in the Balkans.

[2]NATO's AWACS acquisition organization is the NATO Airborne Early Warning & Control (AEW&C) Programme Management Organisation (NAPMO). Thirteen nations are full members of NAPMO: Belgium, Canada, Denmark, Germany, Greece, Italy, Luxembourg, the Netherlands, Norway, Portugal, Spain, Turkey, and the United States. The United Kingdom's participation is limited to attending Board of Directors meetings and other NAPMO committee meetings as required. France attends NAPMO meetings in an observer role, while the three newest NATO members, the Czech Republic, Hungary, and Poland, attend Board of Directors meetings as observers. Iceland does not participate in NAPMO. The NATO AEW&C Programme Management Agency (NAPMA) is the executive agency for NAPMO.

potential for interoperability. Because this study addresses the interoperability of U.S. and NATO allies' air forces, the following discussion focuses on U.S. and NATO AWACS programs, with selected mention of U.K. and French AWACS.

The NATO AWACS fleet has capably met NATO airborne early warning (AEW) mission requirements for almost 20 years. Several factors led to the development of a NATO program based on U.S. AWACS: a common and urgent need existed for air surveillance against a major threat (the former Soviet Union); the U.S. AWACS was the only viable option during the late 1970s; and the program integrator (Boeing) attempted to ensure fair distribution of economic benefits among participating nations.[3]

Today, several factors keep the programs synchronized and interoperable. For example, there is an enduring need for airborne air surveillance and, now, control; in the various modernization programs, there has been an equitable distribution of program cost and industrial benefits; and some formal and informal mechanisms foster interoperability (e.g., common research, development, test, and evaluation [RDT&E] and international working groups). In addition, U.S. Air Force personnel are actively involved in both the acquisition and operations of NATO AWACS, and U.S. and NATO AWACS participate in combined operations and training.

Even with this close cooperation, there are AWACS interoperability challenges. In the next section, we discuss interoperability challenges with regard to systems, missions and operational concepts, political concerns, training, and future U.S. and NATO plans. We then discuss mechanisms to foster AWACS interoperability. Next, we discuss cost implications of the NATO AWACS program. We conclude with observations and suggested actions for the U.S. Air Force.

---

[3]The beginnings of the program were not free from turmoil but were marked by substantial debate, intense negotiations, and political compromise among the Alliance members. Tessmer (1988) provides a detailed account of NATO's decision to acquire its first and only collectively owned and operated defense asset.

## INTEROPERABILITY CHALLENGES

### Systems

Nonsynchronized fielding of AWACS system upgrades can lead to interoperability and fungibility concerns.[4] In many military capabilities, the United States is acknowledged to be ahead of the NATO nations. This is not the case for AWACS; in several ways, the NATO E-3A now leads the U.S. E-3s. The NATO Near-Term Modernization Programme (NMP) (minus the Radar System Improvement Program, or RSIP, which is part of the NMP) brought the E-3As up to approximately U.S. AWACS Block 30/35 capability in December 1997[5]; the U.S. E-3s are still being modified to the Block 30/35 configuration, with completion scheduled in FY 2002.[6] In addition, the RSIP is expected to be on all NATO E-3As by the end of January 2000, while the United States will not complete implementation on U.S. E-3s until FY 2005–2006 (U.S. initial operational capability [IOC] is expected to occur around June 2000). For missions requiring RSIP capability, only a fraction of the U.S. AWACS fleet will be interchangeable with the NATO AWACS fleet until this upgrade is completed.[7]

Moreover, NATO has planned and fully funded additional E-3A upgrades. The NATO Mid-Term Modernisation Programme (MMP) is

---

[4]Fungibility concerns arise because the various AWACS fleets have very similar, if not identical, capabilities and thus are considered not only interoperable but also interchangeable.

[5]Major Block 30/35 upgrades include (1) the addition of an electronic support measures (ESM) system to passively detect, locate, track, and identify emitting air, ground, and maritime targets to improve threat warning and combat identification; (2) replacing the Joint Tactical Information Distribution System (JTIDS) 81Class 1 terminal that uses an Interim Joint Message Standard (IJMS) with a Class 2H terminal that will provide full tactical digital information link (TADIL) J message capabilities (i.e., Link 16–capable terminals), higher data rates, increased interoperability with U.S. and NATO forces, and jam-resistant transmissions; (3) a GPS Integrated Navigation System (GINS) to improve location accuracy of surveillance data; (4) and computer upgrades to support the Block 30/35 modifications. NATO's NMP does not include the GPS upgrade; that will occur during the Mid-Term Modernisation Programme (MMP).

[6]Data regarding the U.S. E-3 schedule are based on discussions with AC2ISRC/C2RS (Caragianis, 1999); by their nature, these data are subject to change.

[7]RSIP is a major AWACS system upgrade, greatly enhancing the operational capability of the radar against smaller airborne targets (i.e., those with low radar cross sections) and improving resistance to electronic countermeasures (ECMs) such as high-power jammers.

under way and will be implemented by December 2004.[8] Meanwhile, the United States is planning to upgrade the computers and displays (including a new tracker, multisensor integration, and datalink infrastructure) on its fleet when funding becomes available, possibly beginning in FY 2004. Moreover, the United Kingdom has not funded upgrades for its E-3Ds beyond NMP and RSIP. Such delays can further exacerbate interoperability and fungibility concerns.

Employment of the same or functionally similar systems does not guarantee interoperability. Development of the basic electronic support measures (ESM) system has been a U.S. and NATO cooperative program. Thus, NATO AWACS use a system similar to that of the U.S. AWACS. The United Kingdom and France have different ESM suites but are negotiating with the United States for the U.S. basic system. However, even if ESM suites are similar, their databases remain a sensitive issue because they are based on each nation's or NATO's intelligence data and thus are not necessarily shared among AWACS fleets without special agreements. During the recent NATO operation in the Balkans (Allied Force), the U.S. E-3s originally used a U.S. database; later they switched to the NATO database when it was offered because it was apparently more accurate and up to date for the theater of operations.

Although communication of tactical intelligence data is common today, the fleet capabilities may be different in the future. Currently, U.S., NATO, and U.K. AWACS can communicate with River Joint, EP-3, and Nimrod; for example, these systems can provide threat warnings or amplifying data on air tracks via Voice Product Net or Link 16. In the near future, however, terminals that receive near-real-time tactical intelligence from U.S. broadcast services will be installed on U.S. AWACS.

---

[8]There are nine MMP enhancements, including the addition of a GPS-integrated navigation system; the addition of two UHF satellite communication terminals for beyond-line-of-sight communications (U.S. AWACS already has SATCOM capabilities); the addition of five situation display consoles (SDCs) to bring the total to 14 (the U.S. AWACS upgraded to 14 consoles during the Block 20/25 program); improved graphical user interface technology at all SDCs and a completely open computing system architecture (i.e., Man-Machine Interface); and integration and fusion of onboard and offboard sensor data and a new tracker to improve tracking and identification (i.e., Multi-Sensor Integration).

Initially, this capability may take the form of a standalone (i.e., not integrated with the AWACS mission software) system such as the Multiple Access Tactical Terminal. Later, this capability (specifically, receipt of the Integrated Broadcast Service [IBS]) will be incorporated in the U.S. version of the Multi-Sensor Integration (MSI) upgrade to the AWACS mission software; thus, the U.S. MSI version will be different from the NATO version developed under the MMP. To maintain fleet fungibility, the other AWACS fleets must obtain these data through other means. Otherwise, they will provide a less complete air picture to aircraft they may be controlling.

Using a "standard" system also does not guarantee interoperability. Link 16 as a standard is a moving target; "compliant" platforms are no longer compliant when the standard is changed. The various AWACS fleets have different implementations of tactical digital information link (TADIL) J message sets because they were implemented at different times. The NATO, U.K., and French AWACS are more alike than U.S. AWACS. The U.S. Navy's F-14 is not Joint Tactical Information Distribution System (JTIDS) fully interoperable with U.S. AWACS and must use voice to communicate certain data. But it is interoperable with AWACS fleets that have implemented backlink; for example, the United Kingdom and NATO have implemented backlink on their AWACS and like this particular enhanced capability, according to NAEWF representatives.[9]

Besides message sets, there is also a need to define and follow standard operating procedures, e.g., the Joint Maritime Tactical Operating Procedures for Link 16 operations, now known as the Joint Multi-TADIL Operating Procedures. However, because of Link 16's wide deployment, any difficulties with this network are the responsibility of a broader community than AWACS. (Link 16 is discussed further in Chapter Nine.)

On a related issue, there are some in NATO who would like to remove the Interim Joint Message Standard (IJMS) message set to get more time slots, but this would blind all of the NATO ground control centers unless a TADIL J translator for these centers is developed. These centers are important to NATO's European nations for Article

---

[9]See Wininger (1999a).

5 operations. Thus, NATO AWACS has a system requirement that other AWACS fleets no longer have.

## Missions and Operations

As discussed in Chapter Five, NATO's past emphasis has been on homeland defense (Article 5 operations), whereas the United States has conducted numerous overseas operations. Thus, the NATO AWACS has operated and trained principally to perform the early-warning surveillance mission, whereas the U.S. AWACS has been called upon to perform both the surveillance and control mission.

The original mission of NATO Airborne Early Warning (NAEW) was to augment ground-based radar systems throughout Europe to counter the low-level threat from aircraft. The emphasis was on surveillance with a limited C2 function. To European nations, "control" was a strategic concern; they wanted control of fighters over their country, i.e., air sovereignty was at stake. However, recent Balkan operations have pulled reluctant NATO nations down the path of airborne control; according to a former NAEWFC commander:

> Whilst strategic radar surveillance remains the basic role of NAEW, a more complex tactical employment of the force involving a mix of air-to-air and air-to-ground control, airspace management, air policing, combat search and rescue, force marshalling and threat warning is becoming the norm in NAEW operations.[10]

The type of control (close, tactical, broadcast, advisory) provided to aircrews depends on equipment limitations, weapon controller workload, and the tactical situation. U.S. AWACS control of U.S. fighters can range across the levels. However, for NATO partners flying less capable, austere fighters (e.g., Greece, Turkey, Italy, Norway, Belgium, and Denmark), close control may be required.[11]

---

[10]See *NATO's Sixteen Nations* (1998).

[11]A mode of control varying from providing vectors to providing complete assistance, including altitude, speed, and heading.

NATO AWACS also has fewer consoles than U.S. AWACS—there are nine consoles on NATO AWACS with two or three for weapon controllers, and there are 14 consoles on U.S. AWACS with four or five for weapon controllers (the United States added five consoles in its Block 20/25 upgrade). In high-intensity conflicts, NATO controllers can be overwhelmed, especially if they are not properly trained. In such cases, the level of control over U.S. aircraft can be reduced; however, "close" control of certain NATO fighters would still be needed.

As a short-term workaround, surveillance consoles can be "converted" to control consoles. The U.S. AWACS can also do this when a maximum effort is needed; in this case, the controllers may be assigned to different functions such as check-in, tankers, offensive counterair (OCA), DCA, or strike. The long-term solution for NATO, of course, is to add five more consoles, which NATO is doing in its MMP.[12]

The introduction of stealth aircraft further complicates airspace control. It is not clear that today's ad hoc procedures using time and space deconfliction are sufficient, especially as the number of friendly stealth platforms proliferates[13] and the emphasis on coalition operations increases (there are U.S. concerns about protecting the characteristics of its stealth aircraft). In such cases, to avoid Blue-on-Blue engagements, the rules of engagement can be made very constraining, but that can increase the likelihood of Red-on-Blue engagements (with Red getting the first shot) because of delays in declaring enemy aircraft as "hostiles." Thus, there is a need to ensure that procedures, or a subset of them, developed for U.S. AWACS to perform the control functions in the presence of stealth aircraft can be provided to NATO AWACS.

---

[12]Although NATO AWACS will get an additional five consoles, the participating nations have *not* stepped up to the additional manning. The United States is already low on NATO AWACS manning without the added consoles.

[13]If not now, at some point in the not-too-distant future it may be necessary for stealth and conventional aircraft to occupy the same airspace.

## Political Concerns

Country sensitivities may lead to interoperability and fungibility concerns. Certain AWACS are not allowed in Turkey or Saudi Arabia, and some countries are sensitive to the fact that AWACS fleets have ESM capabilities, which can be perceived as intelligence collectors.

In high-intensity conflicts in which there are multiple ATOs, non-U.S. AWACS fleets may not be allowed by the United States to receive the U.S. ATO, which can lead to Blue-on-Blue encounters. This can be compensated for by multiple checks with higher-level C2 centers (e.g., CAOC) before declaring an airborne target as a hostile, but such checks can lead to time delays, which can lead in turn to dangerous Red-on-Blue engagements if Blue is not allowed to fire on Red until the last moment. With NATO AWACS flying about 60 percent of all AWACS sorties during the recent Balkan conflict (Allied Force), and with a 70 percent initial operating rate,[14] it would have been difficult to ensure that a U.S. AWACS was available for most U.S. sorties, especially since the bulk of sorties were flown by U.S. aircraft.[15]

## Training

NATO AWACS training has historically focused on activities associated with the traditional homeland defense mission of the Alliance, in which surveillance is assigned a higher priority than control. NATO controllers are therefore not trained to handle large numbers of sorties, reflecting this priority and the correspondingly smaller number of consoles on the aircraft allocated to control.

In addition, U.S. crews have more opportunities for training, especially with fighters, than their NATO counterparts. It can therefore be difficult for NATO AWACS aircrews to achieve and maintain the same level of proficiency as U.S. crews. The NAEWFC is planning to improve its training via modeling and simulation, which should

---

[14]See Proctor (1999).

[15]According to NATO data (Wininger, 1999a), the percentage of AWACS sorties flown by the four AWACS fleets during Operation Allied Force was as follows: NATO (61 percent), U.K. (20 percent), United States (13 percent), and France (6 percent). If total hours on station is used as the metric instead of sortie count, the above percentages remain the same (within 1 percent).

alleviate the problem by diversifying the types of campaigns available for training (e.g., more scenarios with out-of-area operations). A training needs analysis study is now under way at NAEWFC.

Problems associated with inadequate training were apparent during the recent Balkan operation (Allied Force). The mutual lack of familiarity between NATO controllers and U.S. fighter aircrews led to linguistic and procedural confusion. The situation was exacerbated in cases where the NATO AWACS crew were not native English speakers, and whose pronunciation degraded during periods of high activity. Differences in radio terminology were also reported, illustrating a lack of standardization on NATO procedures. All of the above had the potential to create serious misunderstandings in tactical situations.

There are also reports of insufficient training for setting up and operating the new data links. Link 11 has easy and well-defined procedures requiring one day of training to master. Link 16 is more capable and more complex, and thus one day of training is not sufficient. Some of the Link 16 problems that AWACS aircrews encounter are probably due to insufficient training (as with the TADIL J message set implementation, Link 16 training is a larger issue not limited to AWACS).

## Future Prospects

Four European NATO nations are purchasing AEW&C aircraft or are considering that possibility.[16] Greece announced in December 1998 that it intended to buy four Embraer EMB-145s equipped with Ericsson Erieye radars, with deliveries during 2001–2002. Turkey intends to buy four aircraft and to select a winning contractor early in 2000. Italy plans to buy four AEW&C aircraft in 2002, and Spain plans to buy three such aircraft in 2001–2002. There is concern within NATO that these programs will take away from the respective countries' participation in the NATO AWACS program; NATO has indicated that it does not consider these programs to be "in-kind" contributions similar to the U.K.'s E-3D program.

---

[16]See Hewish and Lok (999).

Greece, Turkey, Italy, and Spain have seen the value of interoperability and system commonality through their participation in NATO AWACS. They should be encouraged to develop their national systems in ways that are interoperable with the AWACS fleets at both the technological and operational levels.

There is no urgent need to replace the AWACS airframe in the near term. NATO estimates a retirement date of 2025 based on a U.S. AWACS assessment of airframe sustainability; however, a recently completed NATO study has stated that 2035 is a more likely date.[17] Furthermore, with current flying rates and assuming the aircraft are designed for 60,000 hours of flight time as Boeing states,[18] a much later retirement date, possibly to 2065, can be computed.

NAPMA and NAEWFC are developing a strategic vision beyond MMP. "Initial" results of a strategic review are due in the spring of 2000. This will not be a one-time effort; they expect to develop a "rolling" plan to account for future changes in the military/political environment, operational experience, technology advances, and equipment obsolescence. They are considering two time periods. The first period is when E-3A continues to operate. They will look at the projected lifetime for the airframe to determine the potential for a phase 2 of MMP. They are reviewing previously endorsed requirements whose implementation was deferred for reasons of affordability or technology availability (e.g., further radar and ESM improvements, re-engining, SATURN radio, Link 22). They are also looking at new requirements based on new operational concepts and will respond to externally dictated air traffic management requirements. The second period is post–E-3A. This is currently a low-level investigation that must depend on and reflect the evolution of the major NATO commanders' (SACEUR's and SACLANT's) concepts for follow-on AEW&C capabilities.

The U.S. Air Force also has many concepts for the future that could affect AWACS interoperability. The Air Force is installing the U.S. Navy's Cooperative Engagement Capability (CEC) on a test AWACS. CEC is a real-time sensor fusion system that enables a variety of air

---

[17]See Wininger (1999a).

[18]See Henderson (1990).

defense systems (ground-, sea-, and air-based) on a network to exchange sensor measurement data to create common composite air tracks. CEC involves (1) high-capacity data exchange of detailed (unprocessed and unfiltered) radar and identification friend or foe (IFF) data among network participants via a directional, high-power, jam-resistant distribution system, and (2) fusion of these onboard and offboard sensor data using a common processor. If the United States integrates CEC capabilities on its AWACS and NATO does not, NATO AWACS could, at best, inject near-real-time Link 16 track data rather than raw sensor data into the real-time CEC network, assuming that a Link 16–CEC interface is developed.

The Air Force is also investigating follow-on AWACS platforms such as unmanned aerial vehicles (UAVs) and satellites, principally as sensor platforms. Placing bistatic receivers on UAVs offers the potential of increased coverage against low-RCS (radar cross section) targets. Further, the Air Force is investigating the possibility of migrating the battle management function to ground facilities, which will receive and fuse data from multiple sensors into a single operational picture. This will also provide room for additional sensor growth on AWACS and may provide a transition step to migrating sensors to UAVs and satellites. Such concepts have major implications for AWACS interoperability.

## MECHANISMS TO FOSTER INTEROPERABILITY

Early and sustained emphasis on industrial participation (IP) by contributing nations' industries in the initial acquisition of NATO AWACS, the NMP, and the MMP has been one of the prime mechanisms for encouraging nations' participation, fostering system commonality, and promoting interoperability.

For example, for the MMP that is now under way, 100 percent of each nation's share of the project's costs will be spent in the nation in the form of IP by national industries. More than 70 percent of that spending will be in the form of direct IP (major AWACS subsystem development as well as involvement in production and retrofit activities), with Boeing providing its "best efforts" to distribute direct IP in proportion to a nation's contribution. NAPMA has praised Boeing for its efforts in obtaining qualified European contractors. (An interesting side note is that NAPMA was able to negotiate a fixed-price contract with incentive fee, something that is difficult for the United States to do.)

It is anticipated that overall life cycle costs can be reduced by harmonizing the designs or requirements of all AWACS users (the MMP is fostering such harmonization, particularly with the U.S. fleet). It reduces costs by paying nonrecurring costs once instead of separately for each fleet; it enhances maintainability in that spares and drawings are common among the fleets; and it reduces the time to upgrade each fleet because the bulk of the design work has been done.

Collaborative development and procurement in the modernization programs have led to common systems at lower cost, again promoting interoperability. First, elements of the NATO NMP (minus RSIP) leveraged RDT&E from the U.S. Block 30/35 upgrade. Second, RSIP RDT&E cost sharing is based on aircraft numbers—NATO pays approximately 17/49th and the United States pays approximately 32/49th of the total cost. Of course, the U.S. portion of the NATO AEW modernization program is significant, comprising approximately 41.5 percent based on NMP costs. NATO has an advantage in the RSIP production phase. NATO did a full buy over three years to reduce costs, while the United States is buying kits in small lots for same-year installs, which is less cost-effective. Third, on future upgrades, NATO will not recoup RDT&E costs from the United States when the United States implements upgrades from the NATO MMP. Later, the United States will provide a "better" tracker, the data link infrastructure (DLI) upgrade, and other software upgrades to NATO and not recoup RDT&E costs.

There are also a number of international and NATO forums and working groups, with NATO AWACS and U.S. AWACS participation, that can foster interoperability. The AWACS International Requirements Working Group (IRWG) is an ad hoc working group established to discuss common AWACS requirements and to increase harmonization among the AWACS fleets.

The Multinational AEW Commanders Conference (MACC) brings together representatives from the E-3A, E-3D, E-3F, E-3B/C, and E-2C platform communities. The NAEWFC chairs the MACC. Data link problems in recent operations were discussed at recent meetings.

The AWACS Interoperability Review Group (IORG), established in 1981, is a forum for joint consultation and coordination to ensure that interoperability is established and maintained between E-3 fleets and between those fleets and ACCS and other NATO or national systems potentially assigned to NATO operations.[19] It is cochaired by NAPMA and the primary member hosting the IORG. The primary focus has been on Link 16–related interoperability, i.e., platform implementation of the message formats (IJMS and TADIL J) and JTIDS terminal integration issues (e.g., fighter backlink discussions). The group is also discussing fighter control harmonization (between E-3s, as airborne C2, in relation to air-to-ground missions).

The JTIDS International Configuration Review Board (JIRCRB) is concerned with Link 16 issues that affect many airborne platforms including AWACS. The NAEWFC heads the NATO delegation to this international body. Finally, the NATO C3 Data Link Working Group (DLWG) is a formal group that defines standards for tactical data links; however, it is not involved in implementation.

## COST IMPLICATIONS

In addition to operational benefits, the procurement and operation of NATO AWACS have probably generated cost savings for the United States. The Air Force bought a total of 34 aircraft between 1975 and 1985. NATO bought a total of 18 aircraft in the same time frame. The increase in both the buy rate and the total program quantity associated with the NATO buy probably decreased the costs of the U.S. AWACS. Although the value of the savings is uncertain, the total could easily have reached hundreds of millions of dollars, even with conservative assumptions on the buy rate and learning curve effects.

The United States funded a substantial portion of the cost of NATO AWACS, and thus the savings realized on U.S. aircraft were certainly more than offset by the U.S. share of spending for the program, which exceeded $2 billion. The actual net cost of the program to the United States is difficult to determine. In the absence of the NATO program, the United States might have felt compelled to build and

---

[19] See AWACS Interoperability Review Group (1996).

operate a similarly sized fleet as part of its NATO commitment. As a consequence, it might have had to bear the entire costs of the procurement program rather than the approximately 30 percent[20] share it has paid.

In addition to cooperatively funding the acquisition of the NATO AWACS, most of the NATO partners also contribute personnel and funds to operate the aircraft. The operations and maintenance costs for these aircraft are on the order of $225 million per year, with the United States paying about 41.5 percent.[21] Had the United States been forced to operate these aircraft at its own expense, it would have spent on the order of several billion dollars more than it has had to spend on the NATO program.

Cost savings for modernization have been substantial and should continue for many years to come. For example, the research and development costs of the RSIP have been funded cooperatively, with slightly over one-third of the costs funded by NATO. The net cost savings to the United States are probably on the order of $100 million. Current efforts are under way to harmonize the systems on each of the fleets to reduce the costs of future upgrades. As AWACS will continue to fill a critical mission requirement for decades to come, additional RDT&E efforts for improvements are likely.

Although NATO and U.S. AWACS are following similar upgrade paths, NATO is currently much farther along. NATO AWACS has completed the NMP upgrade and completed the RSIP upgrade at the end of January 2000. U.S. AWACS is not scheduled to complete those upgrades for another five years. Similarly, U.S. AWACS computers and display upgrades will lag NATO's MMP by a number of years. These schedule differences may reduce opportunities for savings, since the buys of common upgrade kits in any given period will be lower than they would be if the schedules were more closely aligned. These smaller buys may result in higher costs for the upgrades.

The cost implications of the NATO AWACS program for the United States have been largely favorable. Allied contributions have resulted

---

[20]The percentage is an estimate based on the cost information available to the authors on the NATO AWACS procurement program.

[21]See Wininger (1999a).

in lower acquisition, operations, and development costs than the United States would have had to pay had it been forced to fund a similar buy of aircraft on its own. These contributions have further lowered the costs of U.S. AWACS.

## OBSERVATIONS AND SUGGESTED ACTIONS

The NATO AEW program is frequently quoted as an outstanding example of allied cooperation and as a visible and viable symbol of Alliance solidarity. Industrial participation has been a major factor in its success and has not reduced the quality of the product.

The procurement of nearly identical AWACS systems ensures a high level of interoperability between the fleets. There are also many forums in which interoperability issues can be addressed. However, materiel solutions can take time to implement because of current funding constraints.

Nonsynchronized fielding of upgrades can introduce interoperability and fungibility concerns. We list three. First, in the air surveillance function, the United States lags NATO in installation of RSIP. Thus, the Air Force should continue to support RSIP installation on U.S. AWACS and to leverage operational lessons learned from NATO experience with RSIP.

Second, NATO AWACS remains primarily an AEW platform. NATO must improve its training (an initiative is now under way) and requires additional consoles on its AWACS fleet (five are funded in the MMP) to become a control platform. However, NATO leads the United States in funding improved computers and displays that will enhance control functions. The Air Force should support NATO efforts to improve the control function (training; tactics, techniques, and procedures [TTP]; and CONOPS),[22] while ensuring that the lag in computers and display upgrades on U.S. AWACS does not create new interoperability problems.

---

[22]The air control procedures employed by U.S. AWACS have proven effective over time under a variety of conditions, including MTW, peace enforcement operations, and punitive raids. Thus, U.S.-developed procedures should serve as a basis for improving interoperability between U.S. and NATO AWACS to address future needs.

Third, fielding IBS on U.S. AWACS (and not on NATO AWACS) will complicate fungibility. The Air Force should determine the impact of IBS installation on U.S. AWACS on the fungibility of the AWACS fleets with those of NATO, the U.K., and France.

Except for IBS, AWACS fleets are likely to remain largely interoperable until after 2025 (when a new platform may be needed) unless the U.S. Air Force and/or DoD implement CEC widely, have different procedures for air surveillance and control of stealth aircraft, separate C2 and sensor functions, or develop other sensor platforms (e.g., UAVs, satellites) for air surveillance.

Chapter Eight

# GROUND SURVEILLANCE AND CONTROL

AWACS is an integral component of how the U.S. Air Force performs airborne air surveillance, controls the airspace, and conducts air-to-air missions. In a similar manner, the Air Force is now looking to JSTARS to perform airborne ground surveillance and attack support control for air-to-ground missions. JSTARS brings to the U.S. military a new tactical capability to detect and, within certain limitations,[1] track moving ground targets as well as to image stopped targets or other fixed targets in adverse weather and under night or day conditions. Its ground moving-target indication (GMTI) radar can surveil a large coverage area,[2] known as the ground radar coverage area (GRCA), at frequent revisit intervals and can also operate in a SAR mode to image smaller areas in the GRCA. Such a capability can support attacks of mobile enemy forces before they reach the main battle area.

JSTARS GMTI and SAR data are sent to the AOC and the Army Corps to provide ground situation awareness and to support ground-attack C2 and targeting. JSTARS is (or will be) netted with other Air Force airborne assets such as AWACS and Rivet Joint (and possibly the Airborne Battlefield Command and Control Center [ABCCC]) through Link 16. Therefore, the basic capabilities that allow JSTARS

---

[1]Current limitations on tracking mobile targets will be reduced with the development and installation of the Radar Technology Insertion Program (RTIP) sensor and an enhanced tracking algorithm, which are part of JSTARS' preplanned product improvement program.

[2]The coverage area is as broad as a corps front and as deep as the location of the second echelon (Air Combat Command, 1997).

to play a central role in air-to-ground battle management and targeting support are (or are programmed to be) in place. Thus, the Air Force, in collaboration with the other services, is determining how these capabilities might best be developed and then exploited in new CONOPS for ground attack missions, such as interdiction, suppression of enemy air defenses (SEAD), close air support (CAS), and theater air and missile defense (TAMD). An example of such a CONOPS for interdiction is discussed in Chapter Eleven.

Several U.S. European allies recognize the value of developing an airborne ground surveillance capability. France has HORIZON,[3] a helicopter-borne GMTI surveillance radar that is flown on the Eurocopter AS 532UL Cougar. Flying at an altitude of 3 km, HORIZON covers an area of 100 km by 80 km and has a maximum radar range of 200 km.[4] Italy has also developed a prototype helicopter-borne GMTI radar known as CRESO.[5] Flying on the Agusta-Bell 412 at the maximum operational altitude of 1.5 km, the radar range is 60–70 km.[6] In June 1999, the United Kingdom selected Raytheon Systems to develop a SAR/GMTI radar known as ASTOR (Airborne Standoff Radar) to be carried on a Bombardier Global Express, a fixed-wing aircraft that can operate at altitudes in excess of 15 km.[7] Like JSTARS, ASTOR can cover a larger area than the two helicopter-borne systems and has a longer endurance.[8] The U.K. plans to procure five systems, with the first entering service in 2005. Figure 8.1 depicts these systems along with JSTARS' modified Boeing 707-300C aircraft.

---

[3]HORIZON is an acronym for Helicoptère d'Observation Radar et d'Investigation de Zone.

[4]See *Periscope* (1996) and Jackson (1999).

[5]CRESO is an acronym for Complesso Radar Eliportato per la Sorveglianza.

[6]See Jackson (1999).

[7]See Morrocco (1999).

[8]In a coalition operation, the helicopter-borne systems could be used for local battlefield surveillance in a pop-up operational mode to minimize their exposure to threats, for filling gaps in the wide-area surveillance coverage provided by fixed-wing systems (i.e., countering terrain masking and foliage), for providing confirmation (as a second sensor), or for augmenting the potentially limited number of fixed-wing systems available for the operation.

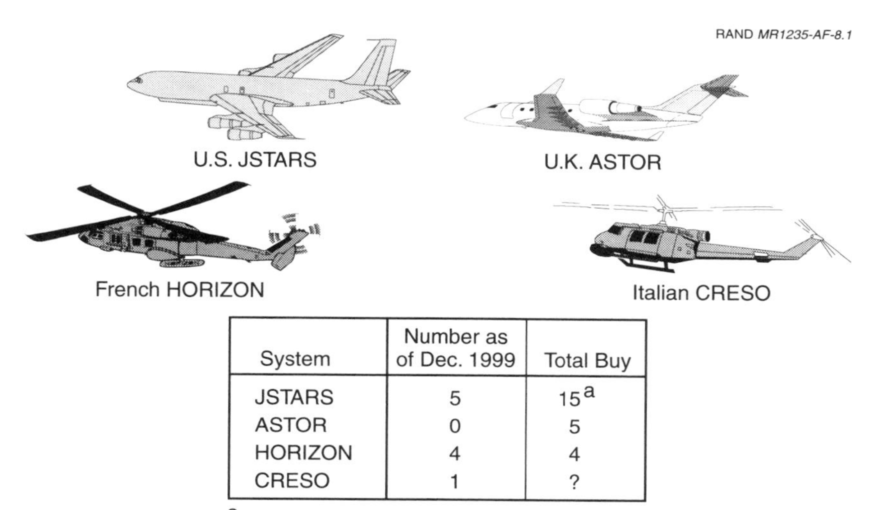


| System | Number as of Dec. 1999 | Total Buy |
|---|---|---|
| JSTARS | 5 | 15[a] |
| ASTOR | 0 | 5 |
| HORIZON | 4 | 4 |
| CRESO | 1 | ? |

[a] The 15th system is in the FY01 budget and has a scheduled delivery date in FY03.

**Figure 8.1—U.S. and European Airborne Ground Surveillance Systems**

NATO has also recognized the value of a ground surveillance capability from airborne platforms and has initiated a program to develop its own capabilities:

> In November 1995, the Conference of NATO Armaments Directors (CNAD) decided that NATO should acquire an Alliance Ground Surveillance (AGS) capability based on a NATO owned and operated core capability, supplemented by interoperable national assets.[9]

As we will see, the NATO AGS program is clearly not following the model of the NATO AWACS program, in which the NAPMO nations decided to acquire a NATO variant of an existing U.S. system, the U.S. AWACS. In the next section, we discuss AGS requirements and then highlight attempts to move beyond concept definition to an AGS program. Next, we discuss other means to achieve interoperability with the airborne ground surveillance systems of our

---

[9]See NATO Consultation, Command, and Control Agency (no date), p. 15.

allies in the event that AGS is not based on a U.S.-developed capability or is not developed at all. We conclude with observations and suggested actions for the Air Force.

## AGS REQUIREMENTS

The current AGS requirements are not definitive.[10] AGS most likely will consist of six to 12 fixed-wing aircraft, enough to maintain one to two orbits, and would operate in a tethered mode (i.e., within line of sight of a ground station) or a nontethered mode (using a limited-bandwidth SATCOM). The AGS would surveil a large area similar in size to the GRCA, but not necessarily with the same revisit interval as JSTARS. One AGS requirements document[11] explicitly mentions the need to "consider" interoperability with certain national ground surveillance systems, specifically the U.S. JSTARS, the U.K. ASTOR, the French HORIZON, and the Italian CRESO.

Discussions continue about the number of onboard consoles. A Northrop Grumman proposal has 11 consoles on an Airbus 321 as compared to JSTARS' 18 consoles on a Boeing 707. ASTOR will have three consoles. SHAPE recommends a minimum of five consoles. Such discussions raise the issue of whether the AGS aircraft will be a surveillance platform, a battle-management platform, or both. For maximum flexibility, the platform should be able to perform both functions depending on the needs of the coalition commander, as with JSTARS. In any event, there will be a common ground station developed as part of the AGS program.[12]

The NC3A has created an AGS testbed to evaluate various AGS options and concepts relating to surveillance, identification, data fusion/data exploitation, interoperability, and integration of AGS data

---

[10]The discussion in this section and the next section on the history of the AGS program and the direction provided by the CNAD is based on visits with Hamp Huckins of the JSTARS International Program, Joe Ross of the NC3A, Robert Bruce of the Office of the Deputy Under Secretary of Defense for International and Commercial Programs, and Pamela Roose of Motorola, in addition to open source material.

[11]See Conference of NATO Armaments Directors (1997).

[12]Some members believe that NATO should collectively finance and develop a common ground station only and leave the expensive sensors in the hands of the nations.

into the NATO C3I structure.[13] Currently, the AGS testbed comprises a number of systems representing all major candidates for components of the NATO AGS capability—specifically, the four platforms mentioned above, a German ground-exploitation and SAR simulation, the Norwegian Mobile Tactical Operations Centre, and a Danish SAR sensor. The testbed is sponsored to a large degree directly by nations, including Denmark, France, Germany, Italy, Norway, the United Kingdom, and the United States.

## AGS CONCEPT DEFINITION

Early in 1997, the United States presented a fast-track offer to NATO, over the objection of the U.S. Air Force, to provide two production versions of JSTARS (E-8C) at a reduced cost (and to be paid back at a later date) for delivery in 2000–2001. The CNAD rejected the offer in November 1997, presumably because there was no urgent need, JSTARS was an expensive solution, there were limited industrial benefits, and advanced technology transfer was minimal. The CNAD then requested fresh concepts. The United States proposed an alternative concept based on the JSTARS RTIP sensor and a platform of NATO's choice to meet the AGS requirement.[14]

The April 1998 CNAD reviewed the fresh concepts, including the new U.S. proposal, and endorsed two concept definition studies—an air-segment study based on the JSTARS RTIP sensor promoted by the United States and four other NATO nations,[15] and a common ground-segment study with the study team led by Germany. A second air-segment study based on a Standoff Surveillance and Target Acquisition Radar (SOSTAR) concept promoted by France, Germany, Italy, and the Netherlands was approved in the fall of 1998. The United Kingdom did not offer a concept and did not participate in the studies because it was in the midst of an acquisition decision on ASTOR.

---

[13]See NATO Consultation, Command, and Control Agency (1999).

[14]For a description of this proposal, see Aldous (1999).

[15]The four NATO nations are Belgium, Canada, Denmark, and Norway. The proposal is now called NATAR (NATO Transatlantic Advanced Radar), and Luxembourg has joined the effort (see Morrocco, 2000).

At the May 1999 CNAD, the results of the three concept definition studies were presented. The CNAD elected not to make any major decisions or to provide strong direction other than to continue the studies. First, the AGS program would enter the project definition phase based on the RTIP sensor. Carrying out this option to its conclusion may be cost prohibitive for U.S. NATO allies (e.g., the recent U.K. ASTOR decision did not select a sensor based on RTIP[16]). Also, there is little allied support for this option without the sharing of RTIP technology. Finally, with allied emphasis on AGS support to the ground commander, there is no agreement on the need for a large aircraft with significant battle management capabilities, a function that can be performed at ground stations.

Second, the SOSTAR group would continue to evaluate its sensor concept. In the near term, this appears not to be a viable option. In the best of circumstances (adequate funding, significant industrial mobilization, and appropriate international agreements), the option would not be timely and most likely would not be available for more than a decade. However, with reduced European defense budgets, the nations are unlikely to support the development of such an expensive, advanced sensor.

Third, the common ground station team, which includes U.S. and European contractors, would continue to refine their concept. This is an important segment to the Europeans because of their interest in providing support to the ground commander and because some nations believe the focus on AGS should be on the ground station and not the sensor. A U.S. contractor (Motorola) that has experience in developing ground stations for sensor platforms (e.g., the U.S. Army's Common Ground Station, the JSTARS Joint Service Workstation, and the new ASTOR workstations—both the onboard console and the offboard workstation) is providing substantial support to the team.

At a future CNAD, the United Kingdom may enter its ASTOR program into the NATO AGS competition; a second possibility is that the U.K. may "declare" ASTOR to NATO as it did with the E-3Ds. It is unlikely that the NATO nations will base the AGS on an ASTOR solution

---

[16]The U.K. selected the Raytheon proposal, which includes a SAR/GMTI radar derived from the ASARS 2 Improvement Program (AIP) employed on the United States' U-2 aircraft (see Morrocco, 1999).

because of concerns over industrial participation, the U.K.'s selection of a U.S. contractor, and the program's greater emphasis on collecting SAR imagery rather than performing GMTI detection and tracking of moving targets. Although the U.K. has rejected the U.S. offer to participate in RTIP development, the U.K. may still buy RTIP as a possible midlife update to ASTOR, thereby increasing the number of more capable theater ground surveillance systems available to a coalition commander.[17]

## OTHER MEANS TO ENSURE INTEROPERABILITY

Because development and procurement of a NATO AGS system is uncertain and, even if acquired, unlikely to follow the NATO AWACS path, the United States should look to other means to ensure interoperability for ground surveillance and control of air-to-ground missions. The ability of JSTARS and U.S. ground facilities to gain access to and exploit GMTI data from a NATO sensor or other European capabilities may be important in areas of operation where visibility is impaired by environmental factors (terrain, foliage, etc.), when JSTARS (or other U.S. GMTI sensors such as the U-2 and the Army's Airborne Reconnaissance Low [ARL]) are not flying, or when multiple sensors are needed to increase confidence in reported information (i.e., confirmation).

One method is to improve interoperability at the ground station segment. The Coalition Aerial Surveillance and Reconnaissance (CAESAR) initiative proposes to improve near-term interoperability among U.S. and European airborne GMTI capabilities (candidate capabilities include those being simulated at the NC3A's AGS testbed). The objectives of CAESAR are to make U.S. and coalition ground surveillance assets interoperable to maximize the military utility of scarce and expensive resources and to enable synergistic use of differing GMTI capabilities, including coordinated mission tasking, planning, and operations.[18] CAESAR was a standalone demo at the

---

[17]The "other" European options—the French and Italian helicopter systems—are not fungible with JSTARS; they are local Army support assets and are not designed for theater surveillance, although they could augment theater assets such as JSTARS or ASTOR.

[18]See Ross (1999a).

Air Force's Joint Expeditionary Force Experiment (JEFX) 99 and is proposed as a five-year, $20 million FY 2000 advanced concept technology demonstration (ACTD).

The initial focus of the ACTD would be on the output of the ground stations, particularly the development of a common data exchange format for transmission of GMTI data. Currently there is no single standard for GMTI reporting, and the problem will increase over time as new GMTI capabilities are developed. At JEFX 99, the NC3A's EX 2.01 data format was used; in the ACTD, the EX format will be enhanced to incorporate new capabilities such as high-range-resolution (HRR) GMTI data. The ACTD builds on an earlier version of the data format that was used at the Paris Interoperability Experiment (PIE) conducted by the NC3A in 1997 with the participation of the United States, France, Italy, the United Kingdom, and the Netherlands. The EX format is also a requirement for the ASTOR program.

In addition to the common data format, other residual products of the proposed CAESAR ACTD include advanced GMTI algorithms (e.g., identification, tracking, prediction, and fusion) and coalition interoperability CONOPS and TTPs for sensor employment (e.g., gap filling, sensor cuing, optimal phasing, and common MTI picture).

CAESAR is an initial step toward ground surveillance interoperability at a time when GMTI resources are scarce. CAESAR will also support the interoperability requirements of the NATO AGS (note that CAESAR is not a substitute for the NATO AGS).

A number of working groups are concerned with the interoperability of sensor systems in general and GMTI systems in particular. The Standing Interoperability and Applications Working Group on Intelligence, Surveillance, and Reconnaissance (SIAR WG) is tasked to define options, study techniques, and recommend solutions for achieving interoperability between NATO ISR systems, both manned and unmanned.[19] This working group reports to Air Group IV, which is responsible for promoting ISR cooperation and standardization for

---

[19]See Air Group IV (1999b).

the NATO air armaments community.[20] The working group has been focusing on imagery interoperability architectures and related standards.[21] At a recent meeting, the SIAR WG was tasked with assembling a technical support team to investigate the need for a NATO GMTI data standard and, if such a need was found, to determine whether to create a new standardization agreement (STANAG) or to modify an existing one.[22]

The Standoff Surveillance and Target Acquisition Systems (SOSTAS) Interoperability Ad Hoc Working Group (SI AHWG) is a recently created working group tasked by Air Group IV to address the interoperability of ISR systems to improve the effectiveness of NATO forces.[23] The working group will address all image collection systems, including GMTI systems,[24] and will provide a report at the end of 2000.

The purpose of the Common GMTI Format Working Group is to define a set of GMTI standards for U.S. producers, such as JSTARS, U-2, Global Hawk, and Discoverer II, to facilitate the transmission, processing, fusion, and display of GMTI data (Boone, 1999). Developing a standard is not, however, a simple process. It concerns data formats and available data rates between the following nodes: (1) the sensor and the platform processing or preprocessing capability, (2) the platform and the ground processing or final processing segment, (3) the ground processing segment and the

---

[20]Air Group IV is one of six subordinate groups that comprise the NATO Air Force Armaments Group (NAFAG). The NAFAG is one of three main armaments groups (the other two are army and navy groups) subordinate to the Conference of National Armaments Directors (CNAD), which in turn reports directly to the NAC.

[21]The working group is responsible for the following NATO STANAGs: NATO Secondary Imagery Format (STANAG 4545), NATO Standard Imagery Library Interface (STANAG 4559), NATO Advanced Data Storage (STANAG 4575), NATO Primary Image Format (STANAG 7023), Air Reconnaissance Tape Recorder Standard (STANAG 7024), and Interoperable Data Links for Imaging Systems (STANAG 7085).

[22]See Nethercott (1999).

[23]See Air Group IV (1999c), and SOSTAS Interoperability Ad Hoc Working Group (1999).

[24]An MTI subgroup was established to (1) investigate the use of MTI products in surveillance, situational awareness, targeting, and tracking, (2) explore the use of MTI products and identify necessary collateral data, (3) define architectures and interfaces needed to handle MTI data, and (4) recommend MTI standards and data formats.

exploitation system, and (4) the exploitation system and the archiving or dissemination system.

The working group expects to define a format—possibly a hybrid format—by the end of December 2000, with formal coordination within the United States and adaptation discussions with the international community scheduled to take place in 2001. The Office of the Secretary of Defense (OSD) has endorsed the outcome of the working group as the DoD standard. The working group is composed primarily of U.S. government and industry representatives; however, the group is also coordinating with the SOSTAS Interoperability Ad Hoc Working Group and the CAESAR initiative. Representatives from these groups, as well as interested parties from other nations, are invited to group meetings.

## OBSERVATIONS AND SUGGESTED ACTIONS

It is not clear when or if there will be a NATO AGS. The program is clearly not following the NATO AWACS history—RTIP is meeting strong resistance, and other options, albeit with more limited capabilities, exist or will exist. If the AGS program proceeds, interoperability may occur only at the output of the ground station segment.

Initiatives such as CAESAR offer the promise of near-term interoperability among coalition airborne GMTI systems through use of a common data standard at the output of the ground station segment. This could also support NATO AGS interoperability needs if interoperability is so defined. Thus, the Air Force should evaluate the results of the CAESAR demonstration at JEFX 99 to determine the value of receiving, exploiting, and fusing ground surveillance information from the ground segments of the different airborne GMTI systems. The Air Force should also investigate the value of receiving European GMTI data on board JSTARS and determine a mechanism for JSTARS to receive such data.

One solution that would lead to greater interoperability lies in the creation of a standard GMTI data format that could be used by all GMTI producers and users (not just the ground station segment) for the transmission, processing, fusion, and display of GMTI data. The Air Force should continue to support the development of a common GMTI data format for U.S. and European GMTI assets.

Finally, the Air Force should help harmonize U.S., NATO, and European allies' GMTI capabilities by supporting the development of a multisensor operational employment concept to ensure that European assets can complement JSTARS.

Chapter Nine

# TACTICAL DATA LINKS

There has long been a need for interoperable data communications for fighter aircraft. Today, most U.S. and NATO allies' fighters communicate using unsecure analog radios that provide only interactive voice communications. This severely limits the coalition partners' ability to reliably share a wide range of combat data in addition to voice over a secure, jam-resistant communications network.

Communications systems that include TADIL capabilities offer a near-term solution for exchanging digital data over a common network that is continuously and automatically updated. Precise quantitative information (data) can be sent faster and more reliably via direct digital (i.e., computer-to-computer) communications. In addition, text messages need only a small fraction of the communications resources that interactive voice messages require and can also be delivered much more reliably than voice in high-stress combat conditions.

Moreover, digital modulation[1] offers many advantages over analog modulation. Four of these are particularly important: the ability to send data; the ability to encrypt voice or data;[2] the use of error detection and correction coding, which increases the reliability and

---

[1]Digital modulation means that information (voice or data) is transmitted as a sequence of discrete symbols, each of which represents a small number of bits. Voice must be converted to a digital stream, a process that is performed by a vocoder.

[2]Analog voice can be scrambled, but this is much less secure than encryption of digital voice and also tends to degrade intelligibility.

quality of transmissions over channels affected by noise, interference, or fading; and, depending on the digital modulation scheme used, a means to distribute energy in ways that can hide the signal to provide for low probability of detection or resistance to jamming.

## TADIL J, JTIDS, AND LINK 16

Several communications systems have been developed over many years to support TADIL communications, or the near-real-time exchange of data among tactical data systems. Each such system is specified by hardware/software characteristics (e.g., waveform, modulation, data rates, transmission media, etc.) as well as by message and protocol standards. The most recent system is the JTIDS/TADIL J system, which is commonly referred to as Link 16 in the United States. Link 16 is an encrypted, jam-resistant, nodeless tactical digital data link network established by JTIDS-compatible communication terminals that transmit and receive data messages in the TADIL J message catalog.

Link 16 data communications standards and technology were developed in the U.S. JTIDS program, which began in 1975. The first JTIDS terminals or Class 1 terminals were large and were installed only on AWACS and at U.S., U.K., and NATO ground-control facilities. Smaller JTIDS terminals (Class 2) were also developed. However, because of their high cost, large size, and reliability issues, only a limited number of such terminals were procured to equip U.S. fighters specifically—U.S. Navy F-14Ds and a single squadron of U.S. Air Force F-15Cs.

The MIDS program was created to put small, lightweight Link 16 terminals on U.S. and participating allies' fighter aircraft. MIDS is a major international program led by the United States, specifically the U.S. Navy, and has a Navy captain as its program manager. By international agreement, the deputy program manager MIDS is a French military officer.[3] The countries funding the development of MIDS are the United States, France, Germany, Italy, and Spain.

---

[3]This management arrangement reflects the cost shares of the international program partners, with the United States and France contributing the largest share of program costs.

With the Low Volume Terminal (LVT) and Fighter Data Link (FDL) terminals—the two terminals being developed and acquired under the MIDS program—Link 16 communications networks will encompass all critical airborne assets involved in air combat, including U.S. F-15, F-16, and F/A-18 aircraft and selected NATO-ally fighters. As indicated in Figure 9.1, MIDS will link fighters to airborne controllers, selected to ISR collection and exploitation centers, and to ground-based C2 nodes such as DCAOCs.

Link 16 can provide a range of combat information in near-real time to U.S. and NATO allies' combat aircraft and C2 centers. The displayed information includes an integrated air picture with both friendly and hostile aircraft locations, general situational awareness data, and amplifying data on air and ground targets, including air defense threats. This will contribute to the integrated control of fighters by either ground-based or airborne controllers and will greatly increase the fighters' situational awareness and ability either to engage targets designated by controllers or to avoid threats, thereby increasing mission effectiveness and reducing fratricide and attrition. An in-depth description of the U.S. Air Force concept of Link 16 employment (COLE) for counterair, interdiction, SEAD, and

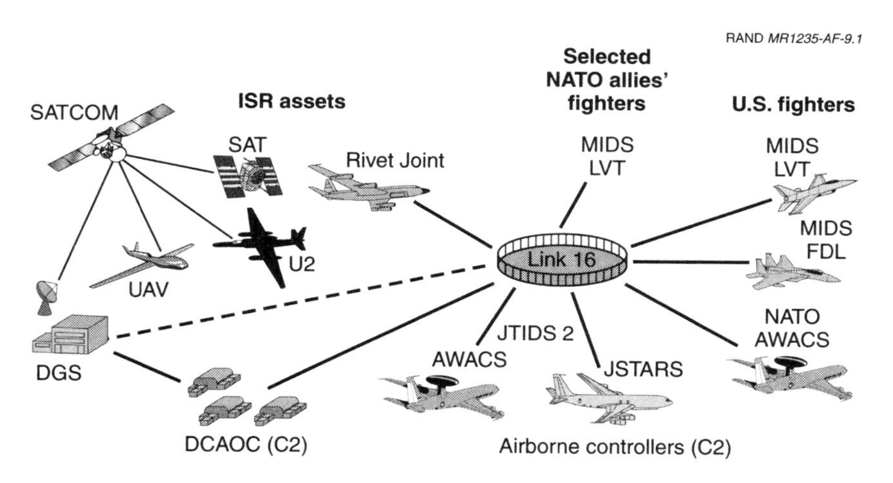


**Figure 9.1—MIDS in the Future Interoperable Tactical Communications Architecture**

CAS missions can be found in the COLE document prepared by the Link 16 System Integration Office.[4] This document describes the information that will be exchanged, how it will be used to support each mission, and the data link architecture that will be employed.

Table 9.1 provides a representative list of the various Link 16 terminals and the platforms (both U.S. and NATO allies) on which they are currently installed or planned for the near future (2010). In principle, if any of these platforms are within line of sight, they could establish tactical communications using Link 16.

## LINK 16 TERMINOLOGY

Because Link 16 terminology is not standardized within the United States or within NATO, we list here the specific standards to clearly indicate how we are using the terms in this report. We also compare U.S. and NATO definitions and standards.

As discussed above, Link 16 uses JTIDS-compatible communication terminals that transmit and receive data messages in the TADIL J message catalog. Specifically, the terminal interface standards (hardware/software) are presented in the JTIDS System Segment Specification (SSS) (DCB79S4000C), and the procedural interface standards (message formats and protocols) are presented in the TADIL J Message Standard (MIL-STD-6016).

These definitions and standards can be illustrated by examining the process for information exchange for a particular mission. Figure 9.2 illustrates this process for the counterair mission. The AWACS surveillance sensor detects a threat. An AWACS crew member prepares the information that will be sent to the F-15C using the situation display console (SDC). The flight processor takes the information and formats it into TADIL J messages. The JTIDS Class 2H terminal encrypts the messages and transmits them to the JTIDS network. The F-15C's JTIDS Class 2 terminal receives the messages, decrypts them, and filters out nonrelevant messages. The flight pro-

---

[4]See Electronic System sCenter (1997).

**Table 9.1**

**Representative Installations of Link 16 Terminals**

| Terminal | Current | Planned (2010) |
|---|---|---|
| JTIDS Class 1 | None | None |
| JTIDS Class 2 | US: F-14D, E-2C, ABCCC, JSTARS, MCE/TAOM, Rivet Joint, F-15C,[a] submarines<br>UK: ADGE, Tornado F3,[b] NIMROD MR | No additional systems |
| JTIDS Class 2H | US: AWACS, MCE/TAOM<br>NATO: AWACS, NADGE<br>UK/FR: AWACS | No additional systems |
| JTIDS Class 2H Shipboard | US: aircraft carriers, destroyers, cruisers | UK: carriers, destroyers |
| JTIDS Class 2M | US: FAAD, Patriot<br>NL/GE: Patriot | No additional systems |
| JTIDS Class 2R (never developed) | None | None |
| SHAR (2R derivative) | None | UK: Sea Harrier |
| MIDS LVT(1) | None | US: F-16, ABL, F/A-18A/F, Navy ships, submarines<br>FR: Rafale, AF ground C2, Navy platforms<br>GE: EF-2000, ACCS platforms, Navy Frigate 124<br>IT: Tornado FBX/SEAD, AMX, EF 2000, Navy platforms<br>SP: EF-2000, EF-18<br>UK: EF-2000, JSF |
| MIDS LVT(2) | None | US: FAAD, THAAD, other C2<br>FR: Army platforms<br>IT: Ground C2 (AF & Army)<br>SP: ACCS platforms (AF) |

**Table 9.1—continued**

| Terminal | Current | Planned (2010) |
|---|---|---|
| MIDS LVT(3)/FDL | None | US: F-15A/E |
| Specific terminal to be determined[c] | | US: F-117, A-10, F-22, B-1, B-2, B-52, JSF<br>UK: JSF |

NOTES:

ABCCC = Airborne Battlefield Command and Control Center (USAF).

ACCS = Air Command and Control System (NATO).

ADGE = Air Defense Ground Environment (U.K.).

FAAD = forward area air defense (U.S. Army).

MCE/TAOM = modular control equipment/tactical air operator module (USAF, USMC).

NADGE = NATO Air Defense Ground Environment.

THAAD = Theater High-Altitude Area Defense (U.S. Army).

[a]Eighteen F-15Cs are equipped with Class 2 terminals.

[b]Three squadrons of Tornado F3s are equipped with Class 2 Link 16 terminals.

[c]At the end of 1999, Air Force data link plans envision incorporating Link 16 terminals on all fighters and bombers. Terminal selection has not been made.

cessor then extracts the content from the messages and displays the information on the F-15C's multipurpose color display (MPCD).

The JTIDS-compliant radio equipment and the TADIL J message formats and protocols are clearly illustrated. The definition of Link 16 provided above includes just these two components. A broader definition of Link 16 is depicted in Figure 9.2. This system-of-systems concept includes the systems used by the aircrews to perform the functions to move the information from one aircrew to another. Although this broader definition is not used in this report, it clearly depicts the aircrews' role in Link 16 and the need for interoperability at the aircrew level.

Within the United States, confusion arises when JTIDS and Link 16 are used interchangeably for the data link. JTIDS and JTIDS-compliant radio equipment (such as MIDS) are just the communications element. There is also confusion surrounding the use of TADIL J. Some want the term to apply to the link, and others want the term to apply only to the message formats and protocols (as

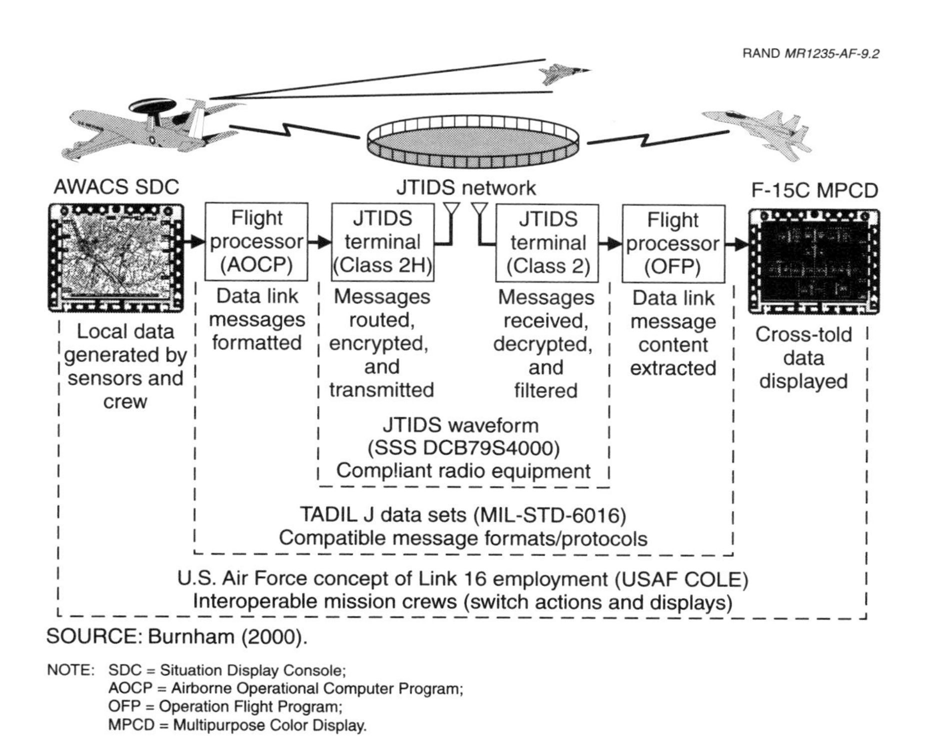


SOURCE: Burnham (2000).

NOTE: SDC = Situation Display Console;
AOCP = Airborne Operational Computer Program;
OFP = Operation Flight Program;
MPCD = Multipurpose Color Display.

**Figure 9.2—Counterair Example of Link 16 (JTIDS/TADIL J) Employment**

defined by MIL-STD-6016). In this report, we use TADIL J only for the message formats and protocols.

NATO has a different view of this terminology. The TADIL J messages and protocols become "Link 16" (STANAG 5516), while the JTIDS communication element becomes "MIDS" (STANAG 4175). Thus, NATO uses Link 16 in a narrower sense than that used in the United States. There are also differences in standard operating procedures: The United States uses the Joint Multi-TADIL Operating Procedures (JMTOP) (Chairman Joint Chiefs of Staff Manual CJCSM 6120.01), and NATO uses Allied Data Publication-16 (ADAP-16). JMTOP has been recommended to NATO for adoption. The different specifications are listed in Table 9.2. It is probably not necessary to

**Table 9.2**

**Link 16/TADIL J/JTIDS/MIDS Specifications[a]**

| Standard | U.S. Specifications | NATO Specifications |
|---|---|---|
| Terminal interface standards (hardware/software) | JTIDS SSS<br>DCB79S4000C | MIDS<br>STANAG 4175 |
| Procedural interface standards (message formats/protocols) | TADIL J<br>MIL-STD-6016 | Link 16<br>STANAG 5516 |
| Standard operating procedures | JMTOP<br>CJCSM 6120.01 | ADATP-16 |

SOURCE: Burnham (2000).

[a]There are other specifications. These are the principal ones.

resolve the differences between the United States and NATO on Link 16 terminology as long as both sides understand these differences.

## SUMMARY OF MIDS CASE STUDY

The MIDS program, which is currently in the final phases of engineering and manufacturing development (EMD), is of high interest to the DoD, as evidenced by Secretary of Defense guidance to the U.S. Air Force to join the program. The international participating nations see it as a successful cooperative program that will provide a near-term solution to a long-standing need for interoperable data communications for fighters. Although the United Kingdom is not part of the MIDS program, it is acquiring another Link 16 terminal known as SHAR (for Sea Harrier) to install on some of its fighters. Thus, six major NATO nations will soon have interoperable, encrypted, jam-resistant communications on their newest fighters. Given the importance of the program for enhancing interoperability with selected NATO allies, MIDS was regarded as a good candidate for a case study.

The MIDS case study, however, is different from the other case studies in which potential solutions to interoperability problems are analyzed and discussed. In this case study, the near-term solution for an interoperable communication system has already been selected, and it is MIDS.[5] Thus, this case study is really an acquisition case study that highlights the programmatic complexities of cooperative initiatives designed to enhance interoperability among coalition forces. The study assesses the advantages and disadvantages of achieving data link interoperability with coalition partners by means of the MIDS program, one of the few international system development programs that has enjoyed sustained international support for an extended period of time.

Below we describe the three major reasons for the MIDS program, summarize our observations of the case study, and present suggested actions the Air Force could take to ensure the success of the program. Because of the complexity of the MIDS program and because there is a separate report on the case study,[6] most of the details are presented in Appendix C. There we examine the goals of the program and the MIDS terminal architectures; discuss programmatic issues, including the history of the program over the last decade; review how MIDS grew out of the original U.S. Air Force–led JTIDS joint-service program; discuss projected costs of MIDS production terminals; and compare those costs to the possible costs of JTIDS Class 2R production terminals if the latter program had proceeded as originally envisioned by the Air Force.

---

[5]One of the drawbacks of MIDS, which is shared by other Link 16 terminals, is an aging system design that takes limited advantage of recent technology developments. This case study does not address the issue of whether this program—or, for that matter, JTIDS—will support all fighter data link needs in future military operations. As discussed in our past work (Hura et al., 1998), additional research on this larger issue is warranted. This case study focuses on short-term solutions to urgent operational requirements. More capable and more technologically advanced data link systems such at the Joint Tactical Radio System (JTRS) are under development by the DoD and may meet the more stressing far-term needs of the services. However, JTRS will not be available in the near term. On the other hand, if the MIDS program can be transitioned into the production phase without major delays, the urgent data link requirements of the MIDS program member nations can be satisfied in the near term.

After this research was completed, additional information regarding enhancements to Link 16 became available. In particular, the U.S. military is investigating enhanced throughput (higher data rates) and dynamic network management for Link 16 (Simkol, 2000). These enhancements would mitigate some of the current shortfalls of Link 16.

[6]See Gonzales et al. (2000).

## Why MIDS?

There are three major reasons for the MIDS program. The first is operational and has already been discussed: to provide interoperable data links between NATO allies' aircraft (fighters, bombers) and air-based, ground-based, and ship-based C2 centers. Because of the position location reporting and identification capabilities of Link 16 terminals, MIDS could provide aircraft IFF information, another desire of the NATO allies. Also, if data could be communicated quickly and accurately by means of a data communications network, it could help overcome language barriers between pilots of different nationalities and thus more effectively integrate the air forces of NATO member nations.

Second, the U.S. NATO allies share a desire for international cooperation and technology sharing with the United States, especially since the United States is viewed as the leader in many military technologies. According to senior DoD officials, the full participation of the U.S. Navy and the U.S. Air Force in the program helped ensure the continued active participation of the international partners in the program. European partners continue to be concerned that competing U.S. system developments will draw funding and resources from the MIDS program and potentially reduce DoD commitment to it. Thus, MIDS serves as a useful test case regarding the feasibility of a truly international system development designed to allow for interoperability among NATO nations.

Finally, although many NATO nations would like a Link 16 capability, they are reluctant to buy JTIDS terminals off the shelf from U.S. industry; European nations want to preserve their own defense development and production industrial base. Since the end of the Cold War, defense spending has declined significantly in Europe as well as in the United States. Thus, budget pressures and European desires to gain access to U.S. military technology led the program partners to favor an international acquisition program that would be a cooperative development effort between U.S. and European defense companies.

## Observations

The cancellation of the JTIDS Class 2R program in 1995 and the decision to join the MIDS program have had cost and schedule implications for the Air Force. The additional cost of procuring the MIDS FDL for the F-15 may be as much as $20 million, but the actual cost is probably much less. There is a strong possibility that the Class 2R program would have encountered significant cost growth (for example, the OSD Cost Analysis Improvement Group (CAIG) concluded at the time that the cost projections for the terminal were too optimistic). Had that been the case, the cost advantage of the Class 2R terminal would have been significantly reduced.

More important for the Air Force has been the delay in acquiring a Link 16 capability for the F-15 aircraft. We estimate this delay to be a minimum of almost two years. This estimate includes only the delays associated with delivery of the terminals, but there may be additional delays caused by possible difficulties in coordinating FDL integration with other avionics upgrades and depot-level maintenance for the F-15 fleet. Furthermore, because the MIDS LVT EMD program has incurred a substantial delay as well, there will also be a minimum delay of nearly three years in the Link 16 IOC for F-16s. As a consequence, the F-16 upgrade program has had to be reprogrammed to adjust for the LVT program delay. The delays in acquiring a Link 16 capability for the F-15 and F-16 aircraft are discussed in Appendix C.

There have been benefits to the decision as well. Air Force participation in the program, initially with the FDL and later with LVT, has helped ensure the continuation of the program. This is important to the United States and its European partners. Now that the Air Force is a major participant, continuation of the program should be assured as long as cost and schedule targets are met.

Furthermore, continued Air Force participation in the MIDS program should bring considerable cost and interoperability benefits. By continuing, the Air Force will encourage allied participation in the program during the production phase, thereby allowing for Link 16 interoperability between U.S. and selected NATO allies' combat aircraft and C2 nodes. The substantial Air Force LVT procurement should drive down terminal costs for the U.S. Navy and possibly for the other countries as well. The Air Force should similarly benefit in that Navy and possible foreign buys should drive down MIDS terminal acquisition costs. In addition, the Air Force should be able to leverage the $650 million investment in technology and terminal design that the other MIDS member nations and the U.S. Navy have made.

Nevertheless, MIDS is a complex international program that could be subject to additional delays and future cost growth. Effective execution of the MIDS program in the production phase will present numerous management challenges to the MIDS International Program Office (IPO), including acquisition management and apportionment of production units to user platforms in three services, quality control, and configuration management. Under the existing EMD management arrangement (see Appendix C), the current senior Air Force officer in the IPO has no officially agreed-upon or assigned duties. Thus, maintaining Air Force insight into this complex program is difficult. To ensure its equities as the largest single buyer of MIDS terminals, the Air Force should be directly involved in defining the management structure for the production phase of MIDS.

Despite the many problems encountered in the turbulent history of the JTIDS and MIDS programs (see Appendix C), MIDS is now an important program for both the U.S. Air Force and the U.S. Navy. It will provide the first extensive deployment of a NATO interoperable Link 16 network to MIDS platforms. Furthermore, it appears that both services now have within their reach a Link 16 data communications terminal that can fit within fighter aircraft and still be affordable.

Finally, it should be noted that the LVT and FDL programs are now closely linked. Therefore, while MIDS holds promise for the Air Force, it also possesses programmatic risks for both the F-16 and F-15 upgrade programs because of the linkage to the avionics upgrade programs of these aircraft. However, if the MIDS program can be managed effectively in the production phase and if MIDS platform integration issues are addressed, U.S. Air Force, Navy, Army, and allied participation in the MIDS program will substantially enhance the interoperability of U.S. and participating NATO allies' forces.

## Suggested Actions

To ensure that the production phase for the LVT and FDL programs is successful, a number of actions should be taken by the Air Force. First, the Air Force should closely monitor the LVT EMD program and verify that it is successfully completed. The Air Force must also verify that the EMD program exit cost criteria are met. The Technical Data Package (TDP)[7] must be completed with sufficient detail to ensure that the production of LVT terminals can be undertaken by multiple U.S. vendors. This action should foster competition in the production phase of the U.S. portion of the program and help ensure that cost and performance objectives are met.

Further, the management structure of the MIDS IPO should be modified in the production phase to provide the Air Force with sufficient visibility into the program and commensurate responsibilities for adequate coordination of the MIDS production program with Air Force fighter and other platform upgrade programs. Three options for doing this should be considered.

The first option would have the smallest impact on the existing management structure. In this case, MIDS would continue to be a U.S. Navy–led program. However, the senior Air Force officer in the IPO would be given a clear set of management responsibilities that would be agreed on by negotiation among the U.S. services. These responsibilities would be recorded in the Joint Memorandum of Agreement (JMOA), now under negotiation for the production phase of the program.

A second option is to create a joint U.S. MIDS program within the IPO without changing the international management structure of the program. The Air Force representative on the MIDS Program Executive Council (PEC) could nominate a senior O-6 Air Force officer to the position of joint program director. This may be possible to

[7]A complete TDP is a critical deliverable of the EMD program—it is essential for ensuring competition and contractor readiness for the U.S. portion of the production phase. The TDP will be owned by the MIDS member nations, so the entire TDP or portions of it could be made available to U.S. contractors. It will not provide a build-to-print blueprint of the EMD terminal. However, it should provide sufficient technical detail to facilitate design and production of system components by contractors not involved in the EMD portion of the program.

accomplish without amending the draft JMOA that is now under negotiation for the production phase program, but it might require the approval of international partners in the IPO. Because we have been unable to gain access to Program Memorandum of Understanding Supplement 2 (PMOU S2), which defines the cost shares and management structure for the EMD program and establishes EMD exit criteria, we do not know whether this option would be acceptable to the other MIDS member nations. Furthermore, we do not know precisely what mechanisms it permits for the possible transition of MIDS to a joint program (as has been suggested by knowledgeable parties).

The third option would be to convert the MIDS program into a true joint-service program. One way to do this is to amend the JMOA that is now under negotiation to explicitly call for rotation of the program director position between the Air Force and the Navy. Again, because we have been unable to gain access to PMOU S2, it is not known whether this option would be acceptable to the other U.S. services and MIDS program member nations.

The factors that need to be considered before choosing a particular option include the total additional cost the Air Force would incur by taking a management leadership role in the MIDS program and the risks the Air Force would incur by not doing so. The overhead costs for managing a joint international program could be substantial, and additional costs may be incurred in moving or consolidating program offices to one central location. However, in the long run the costs and risks could be far higher if the Air Force does nothing to reduce the risks of MIDS terminal production and delivery delays.

Regardless of whether the MIDS IPO management structure is changed significantly, coordination between the MIDS program and Air Force fighter SPOs should be improved. Perhaps the most effective way of doing this is to ensure central Air Force management of MIDS terminal acquisition and integration activities. Currently, the Systems Integration Office (SIO) at the ESC nominally has this responsibility. However, funding reductions have limited the ability of ESC/SIO to carry out the added responsibility as the MIDS program proceeds into the production phase. Consideration should be given to provide this office, or another appropriate organization, with clear terms of reference and sufficient funding to help ensure effective central Air Force management of MIDS terminal acquisition for all platforms in the Air Force.

Finally, another important dimension to the effective operational employment of Link 16 and MIDS lies in the cooperative development and use of concepts of operation. In future coalition operations, U.S. aircraft and C2 nodes may communicate via Link 16 with the aircraft or C2 nodes of other NATO nations, so it is imperative that all coalition partners have a common understanding of and definition for the concept of operations for Link 16, including adequate combined training and exercises.

The Air Force Joint Interoperability of Tactical Command and Control Systems (JINTACCS) office, AC2ISRC/C2PT, is charged with coordinating TADIL J message standardization efforts with NATO partners and with representing Air Force positions in U.S. joint and NATO working groups. The activities of this office are vital in ensuring that Link 16 and MIDS can be used effectively in future coalition operations.

Chapter Ten

# FIGHTERS AND WEAPONS

For much of the Cold War, allied interoperability was enhanced by the predominance of U.S.-designed fighters in the air forces of the NATO allies. This commonality of platforms guaranteed some familiarity and provided a shared experience base for planning and executing operations. The fact that the different air forces were flying the same aircraft ensured an understanding of performance characteristics (e.g., cruise speed and altitude, weapon carriage, signature) and may have enhanced the likelihood that they would be effectively employed (e.g., similar weapon employment concepts, increased fungibility). It may also have provided some advantages in terms of reduced logistical requirements and lower costs associated with operating fewer types of aircraft.

As late as 1980, U.S. designs made up the vast majority of the fighter components of all the NATO allies' air forces except those of France, the United Kingdom, and Portugal. Even in the U.K.'s Royal Air Force, over a third of the fighters were of U.S. origin.

Multinational European aircraft manufacturing efforts in the 1970s and 1980s reduced this dominance among the four largest NATO allies' air forces (see Figure 10.1). Over the next ten years, U.S.-designed aircraft will become a small percentage of NATO fighter fleets as the EF-2000 (Typhoon) comes into service. The lack of system commonality between the U.S. Air Force and the larger NATO allies' air forces, both in their fighters and in the munitions that they carry, is of particular concern in that the larger allies tend to participate most frequently in coalition operations. On the other hand, with the consolidation of the European aerospace industry, there will be

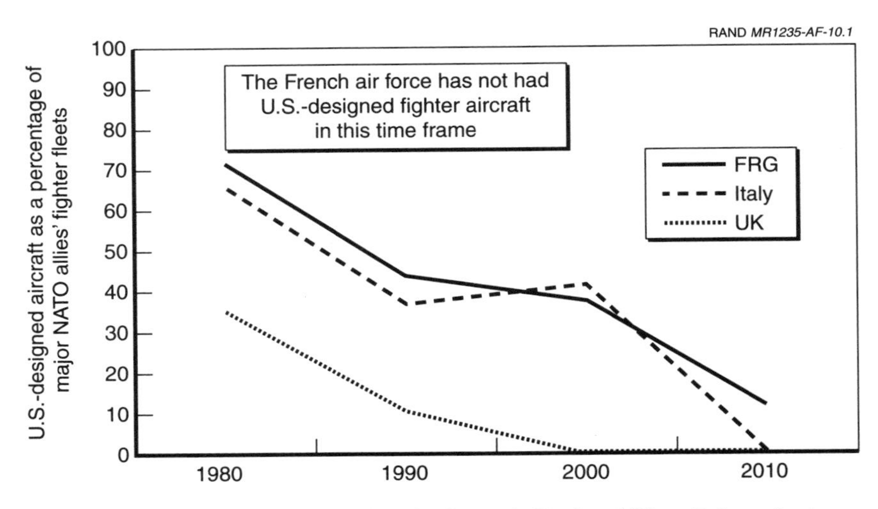


SOURCE: International Institute for Strategic Studies, *Military Balance* (various years) and RAND estimates.

**Figure 10.1—U.S. Fighter Aircraft Becoming Less Common in Major NATO Allies' Air Forces**

fewer types of combat aircraft in the allies' inventories. This trend toward fewer platform types should make it easier to achieve interoperability.

In this case study, we discuss trends in European fighter and weapon systems and their effect on the interoperability of U.S. and NATO allies' air forces. The focus of the discussion is on the ability of these forces to operate effectively in coalition operations. We begin with a discussion of a cooperative fighter development program and then summarize the current and future capabilities of U.S. and NATO allies' air forces. We end with suggested actions for enhancing interoperability.

## COOPERATIVE FIGHTER DEVELOPMENT

In 1996 the United States began to develop the Joint Strike Fighter (JSF). The program grew out of two earlier DoD programs addressing advanced aircraft designs and advanced strike technologies. The

U.K. joined the program as a collaborative partner later that year, and the Netherlands, Denmark, and Norway joined the program as observers in 1997. Italy has been involved as well. In total, the partners and observers are expected to fund approximately 6 percent of the system development cost, as shown in Figure 10.2.[1] The JSF appears to offer promise for enhancing interoperability through the use of a common system. The early interest shown by European nations is indicative of this potential.

However, a preliminary look at the fighter acquisition plans of the NATO nations suggests that this promise is limited. The largest NATO nations have already made substantial investments in other systems. The U.K., Germany, Italy, and Spain have invested approximately $19 billion in developing the EF-2000 and plan to procure

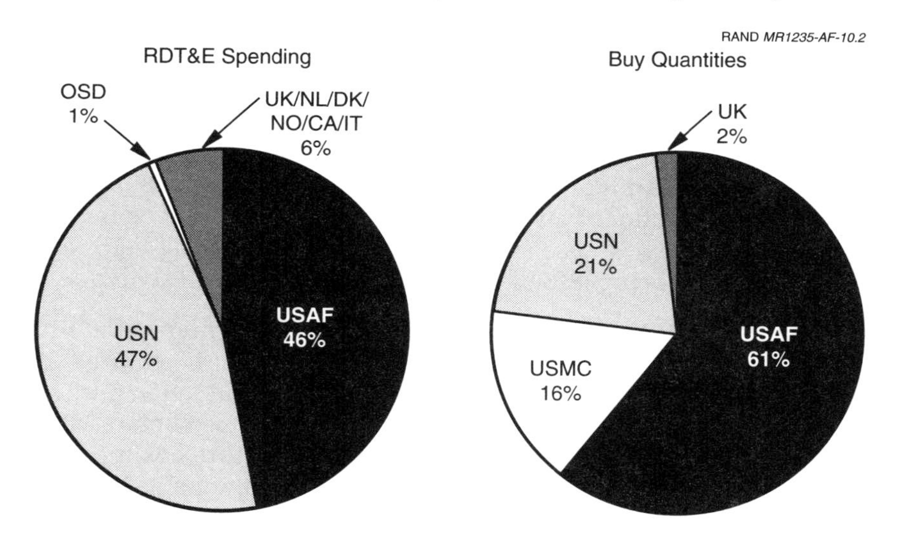


SOURCE: Joint Strike Fighter Program Office (1998).

**Figure 10.2—Allies Playing Limited Role in JSF Program**

---

[1]See Joint Strike Fighter Program Office (1998).

several hundred of them over the next 15 years. France has spent about $9.5 billion developing the Rafale and plans to procure 320 over the next decade.[2]

The U.K. does plan to procure 60 JSFs for its aircraft carriers. That amounts to about 2 percent of the total buy.[3] Italy and Spain could conceivably follow suit, as they also operate ship-based short takeoff/vertical landing (STOVL) aircraft. The small number of ships in this class suggests that the ultimate size of these buys will probably not exceed 200. Thus, prospects for substantial JSF buys for the major European air forces appear limited over the next 10 to 15 years.

The smaller NATO nations (the Netherlands, Belgium, Norway, Canada, etc.) may ultimately buy the JSF, though probably not before 2015. Norway, the Netherlands, and Belgium recently completed the Mid-Life Upgrade (MLU) for their F-16 fleets and will not face a major fighter acquisition decision for the next 10 to 15 years. Plans for the air forces of Poland, Hungary, and the Czech Republic are uncertain. Although the JSF is planned to be affordable, its costs may grow over time. In that case, the relatively small defense budgets of many of these nations may limit their ability to procure even modest numbers of JSFs. Greece and Turkey are still procuring F-16s and thus may not procure JSFs or other new fighters in the next decade.

Although the JSF offers the promise of greater interoperability through system commonality, that promise may not be realized for some years. It is not planned to go into full-rate production until after 2010 and is unlikely to come into widespread use before 2015. The long-term picture is therefore uncertain. Future coalition operations will probably be characterized by less commonality between U.S. and allied fighter forces than has been the case in the past. This lack of commonality may create additional interoperability challenges for planners to address.

---

[2]See Teal Group Corp. (1999a).

[3]After this research was completed, additional information became available suggesting that the U.K. may procure an additional 90 JSFs for the Royal Air Force to replace their Harriers (Braybrock, 2000).

## ALLIED CAPABILITIES

Projections for the fighter components of the air forces of NATO allies in the year 2010 are shown in Table 10.1. Countries were selected for inclusion based on their pattern of participation in coalition operations to date. These estimates are derived from the open-source literature and are thus approximate. They include only fighters that could be deployed operationally.

### Allied Air-Superiority Capabilities

The NATO allies' air forces have an air-to-air capability. Most of the larger air forces have dedicated air-superiority squadrons equipped

**Table 10.1**

**Fighter Aircraft Projections for Selected NATO Allies' Air Forces (Year 2010)**

| Country | Platform | Primary Mission | Number (Combat-Coded) |
|---|---|---|---|
| Belgium | F-16AM | Multirole | 60 |
| Denmark | F-16AM | Multirole | 45 |
| Netherlands | F-16AM | Multirole | 89 |
| Norway | F-16AM | Multirole | 38 |
| U.K. | EF-2000 (Typhoon) | Air superiority | 105 |
| U.K. | Tornado IDS | Ground attack | 84 |
| U.K. | Harrier | Ground attack | 48 |
| Germany | EF-2000 (Typhoon) | Air superiority | 88 |
| Germany | Tornado IDS | Ground attack | 178 |
| Germany | Tornado ECR | SEAD | 35 |
| Italy | EF-2000 (Typhoon) | Air superiority | 59 |
| Italy | Tornado IDS | Ground attack | 45 |
| Italy | Tornado ECR | SEAD | 15 |
| France | Rafale | Multirole | 116 |
| France | Mirage 2000C/N | Multirole | 136 |
| France | Mirage 2000-5 | Multirole | 37 |
| Spain | EF-2000 (Typhoon) | Air superiority | 43 |
| Spain | EF/A-18A | Multirole | 55 |

with modern aircraft and missiles. These capabilities will only increase as the EF-2000 (Typhoon) and Rafale come into service over the next decade. Both aircraft combine signature reduction[4] with advanced avionics. The EF-2000 will first go into service in dedicated air-superiority units. Advanced missiles like the U.K.'s Beyond Visual Range Air-to-Air Missile (BVRAAM) could further advance these capabilities.

The U.K. and France also operate fleets of AWACS aircraft, as does NATO itself. These aircraft are a critical enabler for effective force employment for OCA and DCA missions. The availability of AWACS for peacetime training creates opportunities for NATO allies' air forces to train as they would operate in combat.

## Allied Precision Strike Capabilities

The contribution of NATO allies' air forces in the precision strike mission is more limited. Historically, precision engagement has required that the aircraft be able to find and engage enemy systems autonomously through the use of optical and/or infrared target acquisition systems with laser-guided weapons. This combination allowed aircraft to engage nonmoving targets such as bridges, industrial facilities, or parked vehicles. The NATO allies have traditionally procured relatively few systems designed for this mission; their ability to attack fixed targets with precision strike weapons has been limited.

Although exact numbers are difficult to obtain, even the largest NATO allies' air forces appear to have only a few thousand direct attack guided munitions. However, as a consequence of their experience in Operation Allied Force, most of the allies that took part plan to expand their inventories of precision-guided munitions (PGMs). Long-term plans are uncertain, but recent announcements suggest that allied stocks of direct attack guided munitions will exceed pre-Operation Allied Force levels. These weapons are listed in Table 10.2.

---

[4]Open source reports on the Rafale and EF-2000 note that both aircraft have lower signatures than conventional aircraft such as the Tornado but that neither design is in the same class as the F-117 or F-22.

**Table 10.2**

**Air-to-Ground Precision Munition Capabilities Projections for Selected NATO Allies' Air Forces (Year 2010)**

| Country | Platform | Precision SR Missile | Precision SR Bomb | Precision SO Weapon | Antiradiation Missile (ARM) |
|---|---|---|---|---|---|
| NE | F-16AM (MLU) | Maverick | Paveway II/III | JDAM, JSOW | (—) |
| NO | F-16AM (MLU) | (—) | (—) | (—) | (—) |
| DE | F-16AM (MLU) | (—) | (—) | (—) | (—) |
| UK | Tornado IDS | Brimstone<br>Maverick | Paveway II/III<br>JDAM-like weapon | Storm Shadow | ALARM |
| GE | Tornado IDS | Maverick | Paveway II/III | KEPD Taurus | AGM-88 |
| | Tornado ECR | Kormoran<br>Maverick | Paveway II/III | KEPD Taurus | AGM-88 |
| IT | Tornado IDS | Maverick | Paveway II | (—) | AGM-88 |
| | Tornado ECR | Maverick | Paveway II | (—) | AGM-88 |
| FR | Mirage-D | AS-30L | Matra BGL1000/PW II<br>JDAM-like weapon | Apache SCALP | ARMAT |
| | Mirage-2000N | AS-30L | Matra BGL1000/PW II | (—) | ARMAT |
| | Mirage-2000-5 | AS-30L | Matra BGL1000/PW II<br>JDAM-like weapon | Apache SCALP | ARMAT |
| | Rafale | AS-30L<br>Maverick | Matra BGL1000/PW II | Apache SCALP | ARMAT<br>AGM-88 |
| UK/GE/IT/SP | EF-2000 | (—) | Paveway II/III<br>JDAM-like weapon | Apache SCALP<br>Storm Shadow<br>KEPD Taurus | ALARM |

NOTES: (—) = none planned; SR = short range; SO = standoff; PW = Paveway; ALARM = Air-Launched ARM; AGM = air-to-ground missile; ARMAT = antiradiation missile (French).

The capability to find and engage moving targets with precision weapons was traditionally provided through the use of guided weapons that contained narrow-field-of-view optical or infrared sensors. Although several NATO nations have this capability, few have procured large numbers of these weapons. Thus, NATO allies' ability to use precision strike against moving targets is limited as well.

As noted earlier, collateral damage concerns and the need for increased sortie effectiveness have driven coalition campaign planners to increasingly rely on PGMs. For example, in Operation Desert Storm less than 10 percent of the weapons dropped by U.S. Air Force fighters and bombers were precision guided;[5] in Operation Allied Force, approximately one-third were.[6] In Operation Deliberate Force—a much smaller action—over 90 percent of the munitions used by U.S. fighters were precision guided.[7] There is little reason to suspect that these concerns will be less important in the future.

As a consequence, the ability of the NATO allies' air forces to contribute to future combat operations will depend on the number of aircraft capable of employing precision munitions in their air forces and their inventories of these weapons. Should the allies fail to procure adequate numbers of these systems, their contribution to future operations will be constrained with important implications for burden sharing, especially during long-duration conflicts. This capability shortfall was clearly illustrated in Operation Allied Force. In assessing the lessons learned in Operation Allied Force, U.S. Secretary of Defense William Cohen noted that

> Because few NATO allies could employ precision munitions in sufficient numbers, or at all, the USA conducted the preponderance of the strike sorties during the early stages of the conflict . . . Such disparities in capabilities will seriously affect our ability to operate as an effective alliance over the long term.[8]

Technological developments are creating new opportunities to enhance NATO precision strike capabilities. The advent of GPS-guided

---

[5]See Office of the Secretary of the Air Force (1993).

[6]See USAFE/SE (2000).

[7]Paul Kaminski, Under Secretary of Defense for Acquisition, Technology, and Logistics, quoted in Tirpak (1997).

[8]See Lopez (1999), p. 23.

weapons[9] and offboard targeting systems should create ways to expand the number of platforms that can employ precision weapons at a relatively low cost. GPS-guided weapons do not require the launching aircraft to find the target or guide the weapon. Thus, targeting pods are no longer required. GPS guidance can be used with weapons designed to attack fixed targets (e.g., Joint Direct Attack Munition [JDAM] and Joint Air-to-Surface Standoff Missile [JASSM]) as well as those designed to attack moving vehicles (e.g., Joint Standoff Weapon [JSOW]). At the same time, development of offboard targeting systems has enabled aircraft to engage nonmoving vehicles without acquiring them.

Finally, some of these GPS-guided weapons are considerably less expensive than previous-generation systems, which means that NATO nations may be able to afford them. For example, the current procurement for a JDAM is about $21,000 per unit, representing a substantial decrease from about $77,000 for a GBU-24, a laser-guided bomb with the same warhead.

As NATO nations take advantage of some of these developments, their precision strike capabilities should improve. In recent years, the U.K., Germany, and Italy have put their Tornado strike aircraft through an extensive modernization program that will give them the ability to use both targeting pods and GPS-guided weapons.[10] Similarly, Norway, the Netherlands, and Denmark have invested in the F-16 MLU program to provide the same capabilities. As these modernization programs are completed and additional pods are delivered, the allies' ability to employ precision weapons will increase.

Besides aircraft modernization, some of the NATO nations are increasing the number of precision weapons in their inventories. The U.K., France, and the Netherlands have added to their inventories of laser-guided bombs and other precision weapons in recent years and have decided to procure GPS-guided weapons over the next several

---

[9]Technically, the GPS system does not "guide" the weapon. The information from the GPS signal is used by the weapon to update its inertial navigation system (INS). The weapon's guidance computer then uses the target's coordinates (provided to the weapon before it is launched) and the weapon's current position and velocity (provided by the INS) to develop guidance commands. Often, these weapons are referred to as "GPS-aided INS weapons." Here we use the term "GPS-guided weapon."

[10]See Jackson (1999).

years. These new weapons will increase the number of aircraft able to employ precision weapons against both fixed and stationary mobile targets. The weapons will also greatly enhance their ability to employ weapons in adverse weather, as GPS-guided weapons are relatively unaffected by cloud cover, rain, and other environmental conditions.

However, some difficulties remain. Most of these nations do not have enough targeting pods with optical or infrared sensors and laser designators to equip substantial numbers of the aircraft that are capable had employing them. Further, as of the end of 1999 most NATO nations had not announced plans to procure GPS-guided weapons and thus will not be able to take advantage of their aircraft's abilities to employ such weapons. In addition, these countries lack direct attack weapons containing antiarmor submunitions—such as sensor-fuzed weapons (SFWs)—and have no plans to procure them. These shortfalls may not be a problem in limited operations such as Deliberate Force but could create difficulties in an MTW such as Operation Desert Storm.

**Standoff-Attack Capability.** NATO allies' air forces currently have little in the way of standoff-attack capability. The U.K. recently purchased 61 U.S. Tomahawk cruise missiles, some of which it used in Operation Allied Force. Further, few nations have plans in place to procure standoff weapons. The U.K. and France are cooperating on the development of Apache/Storm Shadow, a new standoff weapon that is in the same class as JASSM. Germany is developing Taurus, which is also in the same class. However, the high cost of these weapons (approximately $1 million) will likely limit their procurement to only a few countries. Even the U.K. and France plan buys of only several hundred weapons each,[11] and the smaller NATO allies may not procure them at all. Further, the British and German weapons are primarily designed to attack fixed targets such as bunkers and runways rather than armored systems. None of them contain advanced antiarmor submunitions. As a consequence, they might not be able to participate in the halt phase of a campaign in a high-threat environment in which standoff antiarmor weapons may be needed to ensure platform survivability.

---

[11]See Teal Group Corp. (1999b).

The shortfall in European standoff weapons does not lend itself to an easy solution. The high cost of these weapons effectively limits their procurement to the larger NATO nations. Even then, the buy quantities are limited and could quickly be used up in a major contingency.

## Allied SEAD Capabilities

NATO allies' capabilities in the SEAD mission area are limited as well. Germany and Italy are the only nations to operate specialized SEAD aircraft. The German and Italian Tornado ECRs are designed to locate, track, and engage enemy air defense systems.[12] Unfortunately, these aircraft are few in number, with just over 50 aircraft between the two countries, and neither nation has plans for further buys. The relatively small size of these units has placed a greater burden on U.S. air forces, which must provide this support in coalition operations. In addition, none of the NATO allies have an electronic jamming platform such as the U.S. Navy's EA-6B. Air forces that lack specialized platforms for SEAD retain some ability to perform the mission at a reduced level of effectiveness. However, campaign planners have shown a preference for using specialized aircraft for this mission, and there is little to suggest that this will not be the case in the future.

Enhancing the allied contribution in the SEAD area may be difficult. Specialized SEAD aircraft are expensive to build and operate, limiting the ability of fiscally constrained NATO allies' air forces to acquire them. One relatively inexpensive way to enhance allied SEAD capabilities would be for the United States to encourage NATO countries operating the F-16 MLU to procure the AN/ASQ-213 High-Speed Anti-Radiation Missile (HARM) Targeting System (HTS). These nations could then dedicate some portion of their fleets to the SEAD mission. Such systems would provide some of the NATO allies' air forces with SEAD capabilities similar to those of the U.S. Air Force at a relatively modest cost (approximately $1 million per unit).

Exporting the HTS pod could conceivably expose the system to greater risk of compromise and exploitation. The desire to enhance

---

[12]See Streetly (1999).

NATO allies' SEAD capabilities would have to be balanced against these risks. Over the long term, SEAD variants of the EF-2000 and/or Rafale could be developed, but they probably would not be available until after 2010.

## U.S. CAPABILITIES

Our review of U.S. fighter force capabilities provides the context for assessing the relative importance of the interoperability enhancements that we describe below. Currently, the U.S. Air Force fighter force includes almost 300 highly capable air-superiority aircraft. Moreover, the introduction of the stealthy F-22 over the next decade will substantially increase U.S. air-superiority capabilities.

U.S. Air Force precision strike capabilities are formidable as well. They consist of almost 400 fighters equipped for all-weather/night precision strike missions and an inventory of tens of thousands of precision weapons. They also include over 100 long-range bombers that can deliver large payloads over intercontinental ranges. Current Air Force precision strike capabilities should also improve as next-generation weapons—such as the Wind-Corrected Munitions Dispenser (WCMD)/SFW, JDAM, JSOW, and JASSM—are brought into service in large numbers.

The Air Force has almost 200 F-16 Block 50 aircraft (F-16CJ) available for the SEAD mission. These aircraft are equipped with the HTS) pod and thus have an enhanced ability to engage adversary air defense systems. These numbers will increase over the next few years as additional aircraft are delivered.

The U.S. Navy also has large numbers of highly capable fighter aircraft. The Navy fighter fleet is made up largely of multirole aircraft. Carrier air wings are currently composed of F-14s and F/A-18s, although this will change as the F-14 is retired and the F/A-18E/F is introduced into service. All these aircraft are precision strike capable. The F/A-18s can employ HARM, JSOW, laser-guided bombs, and a variety of other conventional weapons.

Each of the Navy's 11 aircraft carrier air wings deploys with a total of 48 fighter aircraft. The Navy could deploy carriers to the theater in a short-warning scenario (the number of carriers available in a given contingency is a function of a host of factors, including the location of the conflict and the disposition of the existing fleet when the conflict occurs). The carriers also have four EA-6Bs on board. These aircraft provide a unique standoff jamming capability and can launch HARM missiles as well. However, the Navy may face difficulties in maintaining this capability through 2010 as these aircraft age, and plans for replacing the capability are uncertain. The Navy also has four E-2C aircraft that provide airborne warning and control.

Finally, the U.S. Marine Corps has over 200 F/A-18s with the same capabilities as their Navy counterparts. They often deploy as part of carrier air wings. In addition, they have over 100 AV-8B aircraft for the close air support mission. Small detachments of these aircraft deploy on Navy amphibious assault ships.

Tables 10.3 and 10.4 list the projections for 2010 of U.S. fighter/bomber inventory and the types of precision weapons they can employ. All will be precision strike capable except the F-15C (the A-10 will be limited to Maverick and the F-22 to JDAM). The Air

**Table 10.3**

**U.S. Fighter/Bomber Aircraft (Year 2010)**

| Aircraft | Primary Mission | Total Number (Combat-Coded) |
|---|---|---|
| F-16C/D | Multirole | 600[a] |
| F-16CJ (Block 50) | SEAD | 219 |
| F-15E | Ground attack | 132 |
| F-117 | Ground attack | 36 |
| A-10 | Ground attack | 204 |
| F-15C | Air superiority | 120 |
| F-22 | Air superiority[b] | 171 |
| F/A-18C/D (USN+USMC) | Multirole | 349 |
| F/A-18E/F (USN) | Multirole | 316 |
| B-1 | Ground attack | 70 |
| B-2 | Ground attack | 16 |
| B-52 | Ground attack | 44 |

[a]Only about 252 of these aircraft are equipped with LANTIRN or LANTIRN-like systems and are thus capable of finding and engaging targets autonomously. All 600 can carry GPS-guided weapons.

[b]F-22 will have limited ground attack capability.

**Table 10.4**

**U.S. Air-to-Ground Precision Munitions (Year 2010)**

| Platform | Precision SR Missile | Precision SR Bomb | Precision SO Weapon | ARM |
|---|---|---|---|---|
| F-16C/D | Maverick | Paveway II/III, WCMD | JDAM, JSOW, JASSM | AGM-88 |
| F-15E | Maverick | Paveway II/III, WCMD | JDAM, JSOW, JASSM | |
| F-117 | Maverick | Paveway II/III, WCMD | JDAM, JASSM | |
| A-10 | Maverick | | | |
| F-18A-D | Maverick | Paveway II/III | JDAM, JSOW, JASSM | AGM-88 |
| F-18E/F | Maverick | Paveway II/III | JDAM, JSOW, JASSM | AGM-88 |
| F-22 | | | JDAM | |
| B-1 | | WCMD | JDAM, JSOW, JASSM | |
| B-2 | | WCMD | JDAM, JSOW, JASSM | |
| B-52 | | WCMD | JDAM, JSOW, JASSM | |

NOTE: SR = short range; SO = standoff.

Force fighter force composition will change over the next decade as the F-22 is introduced. In addition, deliveries of the F-16 Block 50 (F-16CJ) will continue at a low rate. Both of these aircraft were incorporated in our 2010 force structure projection.

Deliveries of the JSF will also begin in this time frame, with the first deliveries planned for FY 2007. However, many of the initial aircraft will become part of training and test units and thus would not be part of an operational deployment. A few squadrons may be available in the 2010 time frame. Note that the program is still early in its development, and it may experience delays. A delay of even a year or two could reduce or eliminate the number of squadrons available for operational deployments. Given this uncertainty, we elected not to include the JSF in the tables.

## OBSERVATIONS

The U.S. Air Force has historically enjoyed the advantages of working with NATO allies' air forces equipped with U.S.-designed fighter aircraft. Over the next ten years, the largest of these forces plan to complete their transition to European designs, creating additional interoperability challenges for planners to address.

The U.S. and NATO allies' air forces have substantial air-to-air capability, with the larger nations operating dedicated air-superiority squadrons equipped with modern aircraft and missiles. These aircraft proved more than adequate to meet the limited challenge presented by Yugoslavia during Operation Allied Force. These capabilities should only increase over the next decade as next-generation fighters and weapons are introduced into service.

The precision strike capabilities of these air forces are much more limited. Most of these air forces will soon have large numbers of night attack and precision-strike-capable platforms, but only one will have enough targeting pods to employ these aircraft in this role on a large scale. Relatively modest investments in targeting pods could enhance this capability considerably.

NATO allies' air forces also have limited numbers of advanced strike munitions. In recent conflicts, collateral damage concerns have placed an increasing emphasis on precision munitions, and aircraft survivability concerns have placed an increasing emphasis on standoff weapons. There is little to suggest that this will not be the case in future conflicts. Allied participation in such conflicts could be constrained by a lack of these weapons. This circumstance should improve over the next several years as new European weapons go into production. However, the preference of the larger NATO nations for developing European weapon systems has resulted in higher costs and lower buy quantities than those of comparable U.S. systems, which are already expensive in absolute terms. Thus, the high cost of precision and standoff weapons, be they of European or U.S. origin, may limit the ability of NATO allies to procure significant quantities of these weapons.

SEAD capabilities are similarly limited. Only the German and Italian air forces field a specialized aircraft for this mission. The high cost of these platforms limits the ability of other nations to procure them. A near-term solution might be for the United States to export pod-mounted systems to selected NATO nations for use on their multirole aircraft in SEAD missions. The value of the resulting enhancement to allied SEAD capabilities would have to be weighed against other national security concerns associated with system transfers. Over the long term, dedicated SEAD variants of the EF-2000 and Rafale could be developed, but they are not currently planned.

Enhancing the capabilities of NATO allies' air forces in each of these mission areas will enable them to participate more fully in future coalition operations. Absent these improvements, NATO allies' participation in these missions will continue to be limited, with subsequent implications for burden sharing.

Beyond these mission-oriented enhancements lies a second set of challenges in ensuring that these forces have systems to allow them to work together. This second set of challenges involves systems as well as practices and procedures. Systems that can enable and enhance interoperability include communications, and combat identification. For example, the introduction of tactical data link capabilities, such as the MIDS terminal, should greatly enhance situational awareness of NATO allies' fighter aircraft, allowing for increased mission effectiveness and reduced risk to those systems. This in turn may encourage participation in future coalition operations.

A final challenge lies in developing and practicing procedures so that C3ISR assets can effectively control and manage fighter operations. As the NATO AWACS fleet is enhanced and the AGS is introduced, new procedures may be needed to ensure that these C3ISR assets are able to work effectively with the fighter aircraft of the different NATO nations.

## SUGGESTED ACTIONS

The extensive fighter capabilities of the United States are clearly the "realization" of a key tenet of national military strategy—maintaining the ability to confront adversaries alone if critical interests are at stake. Nonetheless, allied fighter contributions are important, as discussed in Chapter Three and Appendix B. These contributions could be enhanced with improvements to their strike and C3ISR capabilities.

Enhancing allied precision strike and SEAD capabilities will greatly increase the fungibility of NATO allies' air forces, allowing fighters from European nations to substitute for U.S. aircraft in multiple mission areas. This should create opportunities to distribute the burdens of major operations more evenly across NATO members of the coalition. Enhancing allied fighter effectiveness would also provide greater flexibility in allocating forces to ongoing peacetime contingencies. These enhanced capabilities could prove crucial if simultaneous major crises occur.

This review of NATO nations' air forces suggests that the United States should continue to encourage allies to procure more capable air-to-ground targeting and weapon capabilities. The relatively modest costs of the targeting pods and direct attack munitions should put them within the reach of most NATO nations. Further, the United States should continue to encourage its NATO allies to acquire advanced precision munitions. GPS-guided weapons are particularly promising in that they are relatively inexpensive and can be employed without a targeting pod. Large-scale procurement of GPS-guided weapons would enhance NATO allies' fixed-target attack capabilities considerably. The United States should also encourage its allies to procure the targeting pods[13] needed to acquire mobile targets and the weapons needed to engage them.

Encouraging more NATO nations to procure standoff weapons or weapons carrying antiarmor submunitions would probably be more difficult. The high cost of these advanced systems will probably prevent most NATO nations from acquiring them. Even the larger NATO allies' air forces may not be able to procure them in large numbers. Even so, the United States should encourage the allies to acquire standoff weapons to ensure platform survivability in a high-threat environment and standoff antiarmor weapons so they may participate in the halt phase of a campaign.

Enhancing NATO allies' SEAD capabilities may be even more difficult than enhancing their standoff weapon capabilities. The high cost of special mission aircraft limits the ability of most NATO allies' air forces to acquire them. SEAD capabilities could be greatly enhanced in the near term through integration of the HTS pod on F-16 MLU aircraft. Adding this capability would also enhance the fungibility of U.S. and NATO allies' air forces. The benefits and risks associated with exporting the HTS pod would have to be assessed before a decision was reached (such an assessment is beyond the scope of this re-

---

[13]The word "pod" is used for convenience, as it would likely be the basis of any near-term solution. Sensor/designator systems could also be integrated on the aircraft and thus not take the shape of a pod at all.

port). Over the long term, the United States may wish to encourage France and the nations in the EF-2000 consortium to develop SEAD variants of their next-generation aircraft. These new aircraft could then perform more of the SEAD missions in future coalition operations, reducing the burden on U.S. aircraft.

Finally, the United States should continue to encourage NATO nations to acquire systems that enhance situational awareness (through the addition of MIDS, IFF systems, etc.) and improve communications of their fighter fleets. They should also continue to develop practices and procedures to ensure that C3ISR assets such as AWACS and future systems such as AGS are able to work effectively with the NATO allies' fighter forces.

Chapter Eleven

# ILLUSTRATIVE MILITARY VALUE

Previous chapters have highlighted specific interoperability issues based on reviews of past coalition operations and analyses of planning processes and programmatic initiatives that are of prime interest to the Air Force. In this chapter, we examine the implications of these interoperability issues in an end-to-end manner by analyzing representative military operations.[1]

Specifically, we analyze air surveillance during peacekeeping operations, force protection against conventional aircraft and cruise missiles using DCA capabilities, and interdiction of moving columns of armor during the halt phase of an MTW. For each of these operations, we describe an operational concept, identify the system capabilities of the NATO-ally participants, and highlight actual and potential contributions of allied forces in U.S.-allied coalition operations.

## PEACEKEEPING OPERATIONS

In examining the contribution of allies to military coalitions and the benefits of interoperability, the lower end of the spectrum of conflict—peacekeeping operations—must be considered. Because of the lesser strategic risk and value of these operations to the United States, burden sharing becomes an important issue. As illustrated in

[1]To fully measure the military value of interoperability of the United States and its NATO allies in coalition operations, analysis of additional military missions across the spectrum of conflict is needed.

Figure 11.1, we often think in terms of U.S. contributions without due consideration to the contribution of allies.

Allies provide important capabilities that can reduce the burden or fill in for potential shortfalls in U.S. capabilities. Enhancing the interoperability of U.S. and allied systems would increase their ability to do so. In peacekeeping operations, for example, NATO, French, and U.K. AWACS can be used instead of high-demand, low-density U.S. AWACS to provide early warning and air surveillance.[2] A simple calculation indicates that the combined NATO, French, and U.K. AWACS fleets can support about four continuous orbits while the

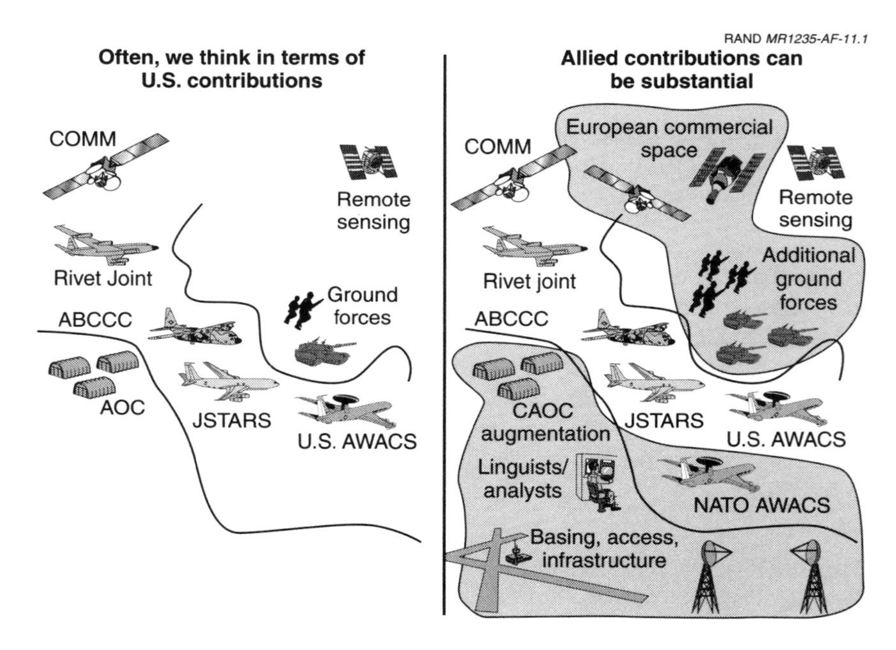


**Figure 11.1—Allied Contributions Are Important in Peacekeeping Operations**

[2]High-demand, low-density aircraft are aircraft that are heavily tasked in support of ongoing operations, but because they are relatively few in number, it is difficult to satisfy the demand.

United States can support about five continuous orbits.[3] This practice proved important during Operation Allied Force, when 87 percent of AWACS sorties were flown by non-U.S. AWACS. U.S. AWACS not only supported Allied Force but were also needed to support SWA and Korean operations and exercises.

NATO partners also provide substantial contribution to CAOCs, including expert personnel who fill key positions and perform critical functions. For example, in Operation Allied Force, allied personnel assumed the battle staff director's position as part of the watch rotation. Allied personnel also manned key CAOC combat plans positions.

U.S. NATO allies can often provide good-quality human intelligence because they are more familiar with countries and regions in which peacekeeping operations have recently taken place. Such a contribution is important in the full range of C3ISR functions, from indications and warning to intelligence preparation of the battlefield to combat assessment.

Allies can also provide space-derived information and services from commercial and government assets. SPOT imagery was used in Operations Desert Shield and Desert Storm and continues to be used in Balkan operations. In fact, the United States purchased a ground terminal to receive SPOT imagery and exploited such imagery to support rehearsal of aircraft missions. Also, imagery from commercial sources was merged with U.S. national imagery to produce a Controlled Image Base product that supported fighter squadrons in the planning of missions in Operation Allied Force.[4]

In addition to air surveillance assets and infrastructure support, perhaps the most important contribution of allies to peacekeeping operations has been in the area of ground troops. In the fall of 1999, for example, NATO allies provided about 85 percent of the 55,000 NATO

---

[3]NATO has 17 AWACS, the U.K. has declared six AWACS for NATO use, and the French have four AWACS, for a total of 27 aircraft. If 75 percent are coded as combat support (i.e., 20 aircraft) and if five aircraft are needed to maintain a continuous orbit, then four orbits can be supported. The United States has 32 AWACS, with 24 coded for combat support; thus, the United States can support about five continuous orbits.

[4]See ERIM International, Inc. (2000), and Electronic Systems Center (2000).

troops serving in the Kosovo peacekeeping operation (Operation Joint Guardian).

## FORCE PROTECTION

One primary mission of the U.S. Air Force is to achieve and maintain air superiority/supremacy by conducting OCA and DCA operations, with an emphasis on OCA operations.

Conducted to attack the enemy's ability to wage an air war, OCA operations include attacks on the enemy's ground infrastructure (e.g., runways, control towers, fuel storage tanks, aircraft on the ground), as well as air-superiority sweep operations that seek out and destroy enemy aircraft over enemy territory. DCA operations counter the enemy's offensive air power and are normally conducted over friendly territory to protect friendly assets. DCA is viewed as an important counterair mission, but it can be less effective than OCA because it is reactive to the enemy's initiative.[5]

Until recently, the air threats have been primarily conventional aircraft (fighters, bombers, and helicopters) and, except for concerns regarding combat identification, U.S. air forces have been well equipped and well trained to conduct effective OCA and DCA operations. In contrast, NATO and its member nations, with their emphasis on Article 5 operations (homeland defense), have developed substantial capabilities to conduct DCA operations against such threats.[6]

However, the proliferation of theater ballistic missiles (TBMs) and, in the future, cruise missiles (CMs) pose significant new security challenges to the United States and to its allies, especially if such missiles are designed with stealth technology or armed with chemical, biological, and nuclear WMD. The United States is active in nonproliferation efforts and is also developing counterproliferation

[5]U.S. air-superiority fighters also conduct "force protection" operations that are designed to protect primary mission aircraft (e.g., ground attack, airlift, and surveillance aircraft) from enemy air attacks. An example of such an operation is fighter escort. In this section, we will focus on DCA operations because of commonalties with NATO capabilities.

[6]They also have well-practiced OCA capabilities against ground aircraft support facilities (e.g., airfields, runways, and fuel and repair facilities), but these are not optimal against modern air defenses.

capabilities such as TAMD. These efforts are meant to deny military benefits to potential adversaries who might develop TBM/CM/WMD capabilities to counter the U.S. ability to conduct military power projection operations.

U.S. military power projection operations often involve (1) forward basing, operational deployments, and the prepositioning of equipment, (2) rapid deployment of decisive force to theater, and (3) precision strike to swiftly meet the defined warfighting objectives and to minimize casualties and collateral damage. Future TBM/CM/WMD threats may counter these tenets of U.S. warfighting strategy by (1) rendering forward forces and equipment vulnerable to attack, (2) slowing and complicating access to ports and bases (politically as well as operationally), (3) creating response dilemmas in the face of casualties, and (4) disrupting and lengthening the conflict. Their ultimate goal is to deny the United States the ability to attain its warfighting objectives and to force the United States to settle for a less desirable outcome. To the extent that NATO adopts, develops, and employs its own military power projection capabilities, NATO forces will also be vulnerable to these future TBM/CM/WMD threats unless it develops its own TAMD capabilities, possibly by leveraging capabilities now being developed by the United States.

In this section, we will explore the capabilities of U.S. and NATO allies' air forces to defend against conventional aircraft (fighters, bombers, and helicopters), CMs, and TBMs to support the collective mission of force protection.

## Against Conventional Aircraft

OCA has been the dominant counterair component used by the U.S. Air Force (e.g., in recent operations such as Desert Storm) to defend against aircraft threats. Today, the Air Force's primary air-superiority fighter is the F-15C, a high-performance, supersonic, all-weather aircraft. In the future, F-22s will conduct air-superiority OCA operations by attacking enemy aircraft over enemy territory with and without air surveillance support from AWACS. At the same time, flights of F-15Cs, F-16s, and JSFs, usually with AWACS support, will conduct OCA and DCA operations. The introduction of the F-22, with its advanced sensors and ability to operate in enemy airspace, has the potential to significantly enhance the situational awareness of less capable air-to-air assets. However, there is no process for disseminating data collected by F-22s to other interested parties (e.g., AWACS, F-15Cs, F-16s, and control and reporting centers and elements [CRCs/CREs]) outside of the flight. Even so, the Air Force's ability to conduct air-to-air OCA and DCA operations against conventional aircraft remains unmatched.

Our NATO allies have a substantial number of air-superiority fighters; however, they were not designed to operate autonomously and primarily perform DCA operations. With NATO's past emphasis on homeland defense and with each nation's concern with air sovereignty, air-superiority fighters performed DCA operations with C2 functions provided by ground control sites located within and operated by each nation. Although the NATO AWACS had limited capability to perform the control function, the aircraft was used primarily to augment the ground-based radar surveillance systems by providing early warning against air threats, especially those at low level.

Our NATO allies are in the process of improving their ability to conduct DCA and OCA operations. They are developing their next-generation air-superiority fighters (EF-2000, Rafale). The NATO AWACS modernization program is in the process of adding consoles for more onboard weapons controllers—this is particularly critical for out-of-area operations where ground control sites either are nonexistent or are not optimally located to perform the control function—and the NAEWFC is improving its training program to ensure that NATO AWACS aircrews are properly trained to perform the control function. The United States and four of its major NATO allies (France, Germany, Italy, and Spain) are developing the MIDS terminal for installation on fighter aircraft, which will allow for encrypted, jam-resistant digital voice/data communications with U.S. and NATO AWACS, both of which are equipped with JTIDS Class 2H terminals. Following the completion of these modernization programs, the capabilities of NATO AWACS and NATO partner air forces to conduct DCA operations against conventional aircraft threats should be close to U.S. capabilities and more than adequate to cope with the threat posed by likely adversaries.

## Against Cruise Missiles

While OCA remains the dominant U.S. Air Force counterair component against conventional aircraft, the emergence of proliferated CM threats, particularly if they are somewhat stealthy or armed with WMD, will likely require a layered defense. The implication for the Air Force and U.S. NATO allies is that enhanced barrier DCA operations will be needed to intercept threats not neutralized by OCA operations.

Barrier DCA operations involve establishing a number of AWACS orbits along the entire threat corridor/border, with each AWACS controlling a number of air-superiority fighters on combat air patrol (CAP). The number of AWACS orbits needed is determined by the length of the threat corridor, the coverage area of the AWACS surveillance radar, the amount of coverage overlap assumed between neighboring AWACS orbits, and the number and performance of the fighter aircraft.

The coverage area of the AWACS radar is dependent on the signature, or RCS, of the threat, with lower-RCS targets yielding smaller coverage areas. Near-term CM threats may be somewhat stealthy with nose/tail signature suppression, whereas more advanced threats may involve all-aspect signature suppression. CMs with nose/tail signature suppression are nominally less of a challenge because their side-aspect signature is similar to that of nonstealthy fighters and can be detected at long ranges unless the CMs approach the AWACS directly, which is unlikely without a well-coordinated attack.[7] CM threats with all-aspect signature suppression require either more AWACS orbits because of their reduced coverage area against the CM or an improvement in the AWACS surveillance radar.

As discussed in Chapter Seven, RSIP is a major AWACS system upgrade that greatly enhances the operational capability of the surveillance radar, especially against lower-signature airborne targets such as CMs. The RSIP capability will provide significant improvement over the current U.S. AWACS radar, reportedly a factor-of-two

---

[7]Even in a well-coordinated CM attack, if air-superiority fighters are forward deployed, they can conduct sweep operations to detect and engage CMs from a side aspect.

improvement in detection range.[8] At the end of January 2000, all NATO AWACS were upgraded with RSIP capabilities; U.S. AWACS are scheduled to complete these upgrades during FY 2005–2006.

We used the following operational concept to assess barrier DCA operations against CMs. Flights of fighters would act autonomously from the AWACS, detecting and engaging approaching CMs. Information about successful engagements as well as about CMs that penetrate the fighter barrier would be passed to the AWACS so that it can direct fighters on CAP to intercept and kill the leakers. The F-22, with its advanced sensor capabilities, has the potential to fill this role; as mentioned earlier, however, it is currently limited in sharing information with other air defense elements. Also, because the F-22 is not designed to operate with EF-2000 and Rafale, it is unlikely that there will be a mix of these fighter types within the specified engagement zone. However, a U.S. and NATO CONOPS to employ air-superiority fighters in such a DCA role needs to be developed.

The AWACS, using either Link 16 (JTIDS/MIDS terminals) or voice communications, will control several flights of U.S. and/or NATO allies' air-superiority fighters (e.g., F-15Cs, F-16s, JSFs, EF-2000s, and Rafales) within its coverage zone and will direct them to intercept approaching CMs. Good situational awareness such as that provided via Link 16 network is essential to the efficient allocation of shooters to targets, airspace control, and adequate deconfliction as well as to minimizing the risk of Blue-on-Blue engagements in such operations. Again, information about successful engagements and about cruise missiles that penetrate the DCA barrier will be passed to any middle and terminal area air defenses (primarily Army and Navy assets).

Because there will be coverage overlap by neighboring AWACS, sensor netting may be needed to avoid dual tracks or loss of track on low-signature targets; further analysis is needed to determine if Link 16 is sufficient or if CEC constructs are more appropriate. If CEC is required—in this example or in other cases in which sensor netting becomes necessary (see below)—and the United States integrates CEC capabilities on its AWACS and NATO does not (as discussed in

---

[8]See Air Combat Command (1996), p. 82.

Chapter Seven), the two fleets will become less interoperable. At best, NATO AWACS could inject near-real-time Link 16 track data rather than raw sensor data into the real-time CEC network, assuming a Link 16–CEC interface is developed.

In addition to the front-line barrier DCA operations, the United States is looking to netted joint-service area air defenses (e.g., Patriot, Aegis) to constitute the middle layer, and terminal area air defenses (e.g., the Close-In Weapon System [CIWS]) to constitute the final layer of a defense-in-depth concept for addressing CM threats. These defense-in-depth operations may be conducted in joint air defense operations/joint engagement zones (JADO/JEZ), requiring close coordination between ground- and ship-based defenses and aircraft to maximize effectiveness and minimize fratricide.

A SIAP that is well integrated with appropriate theater near-real-time tactical intelligence broadcast feeds is essential to help identify, track, and engage threats while minimizing the probability of fratricide. A potential SIAP enabler is the joint composite tracking network (JCTN), a real-time sensor fusion system that enables ship, aircraft, and ground air defense systems to exchange sensor measurement data to create common composite air tracks of fire control system accuracy.

Although the JCTN has yet to be defined, CEC-like constructs are a current model. Also, as mentioned above, interfaces between JCTN- and non-JCTN-equipped participants (e.g., those using a joint data network [JDN] such as Link 16) have yet to be defined. Again, if the United States integrates CEC capabilities or another JCTN system on its AWACS to support such defense-in-depth concepts and NATO does not, the two fleets will become less interoperable.

## Against Theater Ballistic Missiles

As for CM threats, U.S. operational architectures to address TBM threats are typically based on defense-in-depth concepts[9] and can be

[9]In this discussion on force protection, we separated the CM threat and the TBM threat. In actuality, the United States is developing TAMD architectures and operational concepts to address CM and TBM threats together (note that there is usually little discussion of conventional aircraft threats, presumably because the

viewed as consisting of four components: (1) active defenses—terminal defenses (e.g., Patriot PAC-3), midcourse defenses (e.g., Theater High-Altitude Area Defense [THAAD] or Navy Theater-Wide), and boost-phase or ascent-phase defenses (e.g., airborne laser [ABL] or Airborne Interceptor [ABI]); (2) counterforce—air-to-ground systems (e.g., fighters and attack helicopters) and ground-to-ground systems (e.g., Army Tactical Missile System [ATACMS]); (3) passive defenses (e.g., hardened aircraft shelters and revetments, suits and masks to protect against the use of chemical and biological agents, mobility, and camouflage, concealment, and deception [CC&D] techniques); and (4) battle management/command, control, communications, computers, intelligence, surveillance, and reconnaissance (BM/C4ISR) to tie these Army, Navy, and Air Force elements into an effective, integrated, joint-service "system of systems" operational concept.

Other than passive defenses, NATO allies do not have similar capabilities to address TBM threats.[10] NATO can choose to develop such capabilities; however, the TAMD programs listed above—especially the active defenses—represent a substantial investment by the United States, and it is to NATO's advantage to leverage the capabilities now being developed by the United States to the extent that the United States will permit it.[11]

## INTERDICTION DURING THE HALT PHASE OF A MAJOR THEATER WAR

Halt is a mission element within the broad set of air-to-ground interdiction operations (sometimes referred to as battlefield air interdiction [BAI]). The halt phase is particularly challenging because the targets are mobile, hard to kill, and defended by fixed and mobile SAMs and antiaircraft artillery (AAA). Halt analyses highlight a potentially significant divergence between the doctrine, tactics, and ca-

---

United States can successfully engage such threats and thus they are implicitly addressed in TAMD concepts).

[10]The Patriot systems owned by Germany and the Netherlands could be upgraded to the Patriot Advanced Capability-3 (PAC-3) capability, which would provide some defense against TBMs.

[11]Similar observations are made in Gompert et al. (1999).

pabilities of the United States and those of the other NATO nations. Halt brings together several issues, including ground surveillance (e.g., U.S. JSTARS, NATO AGS, or other European nations' airborne surveillance capabilities), dynamic battle management (e.g., airborne versus ground-based), the availability of digital data links (e.g., JTIDS/MIDS terminals), and the employment of standoff precision strike with smart, lethal weapons, especially those carrying antiarmor submunitions.

This discussion focuses first on engagement-level operational concepts and systems needed to interdict moving armor. We then investigate the military value of U.S. and allied air power interoperability in interdiction operations within the larger context of the early halt phase of an air campaign.

## Engagement-Level Considerations

The emergence of more mobile ground forces and more capable air defenses, and the desire to improve efficiency (more kills per sortie) and minimize collateral damage in interdiction operations, are forcing changes in the way interdiction operations will be conducted in the future. In particular, there is a need for new weapon systems and new operational concepts. This section discusses these issues as a necessary prelude to our analysis in the next section of the military value of U.S. and allied air power interoperability in interdiction operations within the larger context of the early halt phase of an air campaign.

**Challenging Targets and Threats.** Traditionally, most air-to-ground missions have been planned one or two days in advance and executed through the ATO. The strike aircraft would use their onboard sensors to acquire the targets and then release unguided or guided direct attack munitions. This is adequate against strategic targets and fixed interdiction targets such as garrisons or lines of communication (LOCs) but is not an effective or efficient method for attacking short-dwell or moving targets. Often, moving targets were targets of opportunity and, if found by the strike aircraft, would be attacked instead of, or in addition to, the fixed targets the aircraft were directed to strike. If aircraft were directed to find and attack moving targets, especially with limited advance knowledge, the operation would devolve to an "armed reconnaissance" operation that may be effective if targets were found and engaged but ineffective and inefficient if targets were not found.

Short-dwell and moving targets are assuming greater importance. For example, rapidly finding and attacking "target-rich" arrays such as advancing armor columns in a halt operation or "sparse targets" such as C2 vehicles, advanced SAMs, or TBM transporter-erector-launchers (TELs) are important future challenges for the U.S. Air Force.

Unfortunately, other adverse threat trends, particularly mobile long-range air defense systems (e.g., SA-10s, SA-12s) and mobile forward-area terminal air defenses (e.g., 2S6 vehicles, shoulder-fired infrared SAMs) further complicate matters. These defenses are difficult to suppress and, because of their lethality, limit the effective employment (i.e., with low attrition) of direct-attack precision weapons. Terminal defenses may force aircraft to higher altitudes where weapon performance may be degraded. Long-range defenses may force aircraft to employ standoff weapons. In addition, if an armed reconnaissance approach is used to find and engage moving targets, the exposure of such aircraft to these advanced air defenses will be increased with a proportional increase in risk to the aircraft.

**System Improvements.** To address enhanced air defenses and mobile target sets, efforts have been under way in the United States for some time to develop high-altitude delivered, direct-attack precision weapons (e.g., WCMD) and effective standoff precision weapons (e.g., JSOW)—particularly variants capable of hitting and killing armored vehicles (e.g., WCMD and JSOW carrying BLU-108 submunitions). WCMD and JSOW variants carrying combined-effects bomblets (CEBs) designed for soft-area targets can be used against the softer (i.e., nonarmor) elements of an armor column, such as trucks, mobile command posts, and air defense units. Note that the emphasis on developing standoff precision weapons is intended not only to ensure safe (minimum risk of aircraft attrition) and efficient (high kills per sortie) prosecution of targets (whether fixed or mobile) but also to minimize collateral damage, given that the rules of engagement have become increasingly stringent.

In addition to its investments in standoff precision weapons, the United States is making investments in offboard targeting assets (e.g., JSTARS improvements such as RTIP, enhanced tracking algorithms, and implementation of JTIDS attack support messages) to support shooters with timely moving-target information. Although standoff is generally a useful weapon characteristic, most current long-range weapons can be difficult to employ effectively against moving targets because they cannot be updated in flight. Specifically, if the targets change direction or speed unexpectedly, the targeting solutions provided by the offboard sensor may be in error, and the weapons may not be delivered to the right locations.

Our NATO partners are not making comparable investments in offboard targeting assets and standoff precision-guided weapons (as discussed in Chapters Eight and Ten, respectively) and instead rely on their SEAD capabilities and shooter self-protection capabilities to allow for the employment of direct-attack weapons. Thus, U.S. efforts in attacking ground targets, particularly nonemitting mobile targets, are substantially ahead of those of its NATO partners.

**Emerging Operational Concepts.** In addition to the development of specific systems, effective future air-to-ground operations against moving targets (e.g., armored fighting vehicles [AFVs]) will require substantial development of operational concepts and associated tactics and doctrine. As U.S. AWACS is to air-to-air missions, JSTARS is potentially to air-to-ground missions; thus, it is natural to investigate the central role JSTARS, a future NATO AGS system, or the U.K.'s ASTOR might play in interdiction missions. Equipped with wide-area-coverage GMTI radars, these systems can bring to NATO and coalition operations a new capability for detecting moving ground targets and—with planned enhancements such as an upgraded radar and an improved tracking algorithm—the opportunity to develop targeting, control, and BM capabilities for air-to-ground missions.

A notional operational concept for interdicting columns of moving armor is depicted in Figure 11.2. The concept begins with target detection (and classification, where feasible), in this case by JSTARS. Then, JSTARS operators track the threat arrays and develop and provide targeting information to coalition fighters, assumed to be on CAP in the general area. In parallel, JSTARS controllers, in contact with the fighters, provide the battle staff on JSTARS with force status information to coordinate attack plans and provide the fighters with

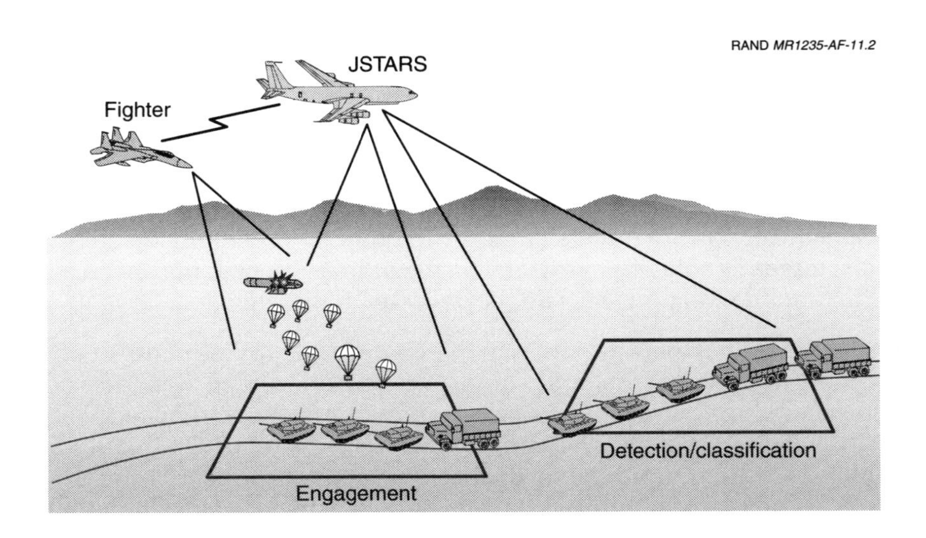


**Figure 11.2—Operational Concept for Interdiction of Moving Armor**

the necessary situational awareness to ensure their safety and effectiveness. Communications with the fighters will be done by voice and by digital data link (e.g., JTIDS/MIDS terminals).[12]

Finally, U.S. fighters will release their standoff weapons (e.g., JSOW) or high-altitude delivered munitions (e.g., WCMD) to achieve the planned time-on-target against the predicted aim points, accounting for uncertainties in the last sensor update, anticipated target motion, transmission of target data,[13] downloading the data into the weapons, and weapon time-of-flight delays. In this form of targeting,

---

[12]Because of the current debate on the future NATO AGS discussed in Chapter Eight, we recommend that future research examine an alternative CONOPS for interdiction missions in future coalition operations. This CONOPS would consist of a variety of GMTI sensor platforms providing GMTI data to ground nodes where personnel would analyze the data and provide targeting information and direction to fighters.

[13]Target data provided to the pilots by JSTARS include accurate aimpoint geocoordinates, weapon-approach azimuth for effective munitions dispensing, and desired time-on-target.

U.S. fighters do not have to acquire the targets with their onboard sensors.[14]

As depicted in Figure 11.2, armored vehicles often travel in columns on roads. In open terrain, however, the armored vehicles could travel off-road, thereby complicating the targeting solution by making direction of movement more difficult to predict. Also, for long-time-of-flight weapons, the predicted aim point is more uncertain. In both cases, uncertainties in the targeting solution can result in reduced effectiveness for weapons with small footprints or when attacking targets moving in short columns.[15]

Because the allies' fighters employ relatively low-altitude direct attack weapons (e.g., Mavericks), JSTARS provides less targeting support than it does to fighters employing standoff weapons. JSTARS can be used to deconflict fighter attacks by directing them to different columns or different segments of long columns. The fighters can then acquire and target the moving targets using onboard sensors. Thus, the targeting solution is not affected by target motion to the degree it is for standoff weapons using offboard-generated data. However, the fighter is at much greater risk to terminal air defenses.

**Weapon Effectiveness Using Offboard Targeting Data.** As stated above, the effectiveness of weapons targeted against moving targets using offboard data are sensitive to time delays (i.e., weapon time of flight, ISR and C2 delays in relaying targeting solution to shooters), target characteristics (e.g., spacing of vehicles, on-road versus off-road movement), and uncertainties in surveillance sensor measurements (current target position and velocity). Figure 11.3 illustrates the sensitivity to on-road versus off-road movement and weapon time-of-flight.[16] The figure shows the armor killing effectiveness in terms of expected kills per weapon for WCMD and JSOW, each deliv-

---

[14]As discussed earlier, the acquisition of moving targets with onboard sensors is still a predominant practice in interdiction operations conducted today.

[15]If there are a large number of targets in a long column traveling along the road, an accurate targeting solution is not critical, as the weapons can be patterned along the road (with some overlap of footprints). The lead and trailing elements may not be attacked because of errors in the targeting solution, but the middle elements will be.

[16]Data for this figure were obtained from two recent RAND studies: Ochmanek et al. (1998) and Rhodes and Harshberger (1998).

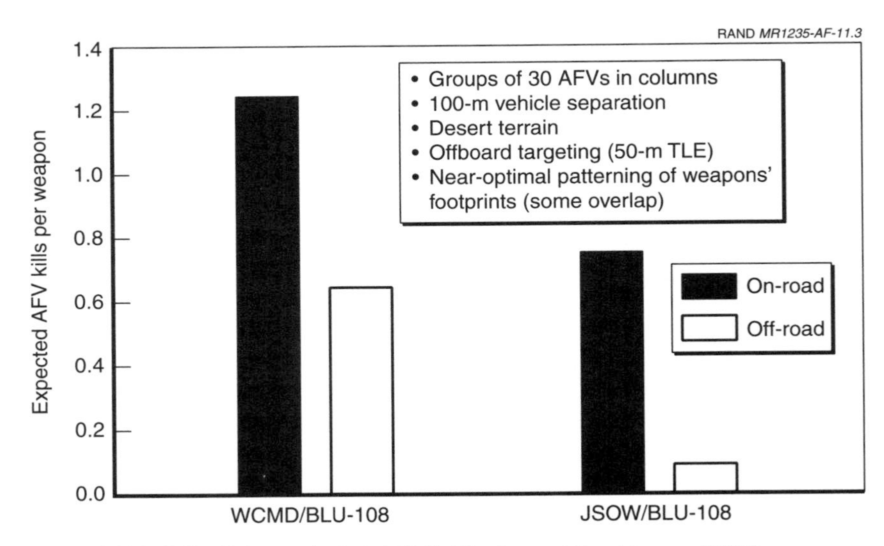


SOURCES: Ochmanek et al. (1998), Rhodes and Harshberger (1998).

**Figure 11.3—Weapon Effectiveness Using Offboard Targeting Data**

ering antiarmor submunitions (e.g., BLU-108s) from high altitude against columns of AFVs advancing either on-road or off-road. In the on-road case, we assume that accurate road maps are available and that the columns do not change their direction of movement when they encounter road intersections.

Whether the targets are on-road or off-road, WCMD is more effective than JSOW because it dispenses ten BLU-108 submunitions, whereas JSOW dispenses only six over a similar-size footprint. Because JSOW has a longer time-of-flight than WCMD, its effectiveness is more sensitive to off-road movement by the armor columns—to an extent that employment of JSOW may not be practical. An operational solution is to emphasize SEAD missions so that standoff weapons are not required. Another option is to improve standoff weapons—e.g., by providing them with the capability to receive in-flight target updates, increasing their submunitions' footprints and adding improved seekers, or developing submunition delivery vehicles equipped with a smart seeker to locate target clusters before releasing submunitions.[17]

We did not assess the weapon effectiveness of the WCMD and JSOW carrying CEBs against trucks and other soft targets in the armor columns because we explicitly modeled attacks only against AFVs in our mission-level analysis. However, we did assume that the weapon effectiveness against these soft targets is at least as good as if not better than that depicted in Figure 11.3. As a consequence, our daily JSOW and WCMD allocations (50 percent to BLU-108 variants and 50 percent to CEB variants) were made in proportion to the composition of the armor columns (50 percent armor targets and 50 percent nonarmor targets).

Offboard targeting data are not used to support the employment of low-altitude direct attack weapons; the targeting solution is developed by the fighter and its onboard sensors. In the case of Maverick, the seeker on the weapon itself locks onto the target before the fighter releases the weapon. For our analysis, we used a single-shot probability of kill of 0.5 against AFVs;[18] the same value was used for both allied and U.S. fighter employment of direct attack weapons.

When used in our mission-level analysis, the kill probabilities above are adjusted to account for a number of operational degrades. Sorties will be aborted because of bad weather, they will be provided with inaccurate information, or they will be mistakenly directed to targets that do not match the weapons being delivered. Over time, as the target set is attacked and destroyed, it will be more difficult to distinguish unattacked targets from those that have been damaged and the columns will not be as uniform or dense, with a resulting decrease in weapon kill efficiency for weapons dispensing submunitions over a given footprint. Because we use a leading-edge attack strategy to counter the enemy force's penetration into friendly territory, there is a greater likelihood that unattacked targets will be among damaged targets, which will confuse the sensors on the BLU-108 submunitions and result in further degrades to overall kill probabilities.

---

[17]The U.S. Air Force initiative to develop a Low-Cost Anti-Armor System (LOCAAS) is pursuing this last alternative.

[18]See Ochmanek et al. (1998).

## Mission-Level Analysis

The discussion above focused on engagement-level operational concepts and systems needed to conduct an interdiction mission against moving armor. In this section, we investigate the military value of U.S. and allied air power interoperability in interdiction operations within the larger context of the early halt phase of an air campaign supporting a notional out-of-area operation in open, desert terrain, such as in SWA, during the 2010 time frame.

The analysis will highlight a diverse set of interoperability issues—ones that span strategic (e.g., deployment and basing) as well as tactical (e.g., weapons and delivery tactics) levels.

**Measure of Effectiveness.** The number of AFVs is widely used in the U.S. defense community as a measure of a ground force's combat potential. Hence, the yardstick used here to measure the relative importance of selected interoperability issues is the number of days needed to halt an invading army,[19] where the "halt" is defined as that time when a specific fraction of AFVs are stopped.[20] This fraction varies but is typically on the order of 50 percent. We used 50 percent.

There is also value in attacking softer targets such as mobile command posts (if they can be identified) to degrade the enemy's command and control of the invading army and trucks that carry personnel, spare parts, fuel, and other consumables vital to the invading army. However, the connection between damage to these support assets and the invading force's combat capabilities is difficult to assess. Nevertheless, we chose to attack these softer targets with CEB variants of JSOW and WCMD because we believe that it is part of a valid attack strategy under the conditions postulated in this analysis. This attack strategy will consume resources (i.e., sorties will be allocated to these targets), and because we did not include any measur-

---

[19]Halt time can be related to the maximum distance traveled by the invading army. For comparative purposes this distance can be correlated to critical potential objectives, e.g., the enemy force's reaching a key city such as Dhahran in Saudi Arabia. Thus, maximum penetration distance is another metric besides halt time to demonstrate the importance of quickly stopping the enemy's advance. We did not use this metric in our analysis.

[20]Stopped AFVs include AFVs killed but also include other elements remaining in a unit that is considered no longer militarily effective.

able value to these attacks in halting the armor columns, the results of our interdiction analysis should be considered conservative.

Similarly, there is often merit in attacking LOCs (such as bridges and other choke points) to slow or stop the advance of an invading army. But the value of such attacks can be difficult to assess, and depends, for example, on the number of alternative roads, the amount of open terrain, and the enemy's ability to quickly repair any damage. In our analysis, we did not include any value to attacking LOCs because we assumed open, desert terrain and the availability of alternative avenues of approach.

**Notional Threat Disposition.** We consider a five-division enemy ground force of heavy armor advancing on two axes, in a two-column formation on each axis.[21] Vehicles are spaced approximately 100 meters apart, with a nominal advance speed of 60 km per day, although this will vary depending on circumstances. Organic to these divisions are a variety of systems, including tanks, infantry fighting vehicles (IFVs), artillery, air defenses, helicopters, and trucks. The total may number approximately 1000–1200 individual systems per division. Of that total, about 600 systems (i.e., about 50 percent of the systems) are AFVs (tanks and IFVs). We explicitly model only the length of time required to stop the AFVs; however, our weapon employment strategy includes delivery of munitions that are effective against both armor and nonarmor targets.

As noted, there are air defenses organic to the advancing columns that pose a threat to the attacking air forces. These defenses are expected to include the Russian-built 2S6 (radar-directed AAA and SAM combination), the SA-15 (radar-directed SAM), and the SA-18 (infrared-guided SAM). In addition, in some cases the formidable SA-12 may be present. If so, we postulate that this SAM would most likely advance with the enemy columns in a leapfrog deployment, possibly some distance back from the main columns at any given time but still capable of significant forward "reach" because of its very long range. Although we did not quantitatively examine the

---

[21]Although the enemy may have ten divisions for an invasion, we envision a short-warning scenario in which five divisions are appropriately positioned and postured at the start of hostilities.

impact of the SA-12's presence, we discuss its impact in qualitative terms.

**Relative Contribution of Forces.** Earlier chapters provided a perspective on the NATO allies' air power contributions to recent conflicts and highlighted the fact that their contributions of fighter aircraft varied substantially. In the one MTW (Operation Desert Storm) in which the United States and its NATO allies participated, the United States contributed roughly 25 percent of its fighter forces, and four major NATO allies made the following contributions: U.K., 28 percent; France 15 percent; Italy 5 percent; and Canada 23 percent.[22] Therefore, for this analysis, we postulate that each country might reasonably supply a similar percentage of fighters to a future conflict of similar scale.

In the case of the U.S. Air Force, we primarily used a slightly modified form of the force deployment described in Ochmanek et al. (1998), which is consistent with historical data. This force is constructed to first support the "enabling" portion of the halt phase (e.g., rear-area asset protection, SEAD and air superiority, and disruption of enemy C3 and transportation networks) and then to support attacks on enemy armor columns. In place, prior to the outbreak of open hostilities, are two squadrons of F-15Cs and a single squadron each of F-16CGs (with LANTIRN), F-16CJs (with HTS), and A-10s. The analysis also assumes that a Navy aircraft carrier is in theater as hostilities commence, resulting in the immediate availability of four F/A-18 squadrons (48 total combat-capable aircraft). When the "enabling" phase transitions to the armor column attack phase, force deployment emphasis shifts to precision-strike-capable aircraft (e.g., F-15E, F-16CG).

This analysis relies on a slightly modified and augmented bomber force compared to that described in Ochmanek et al. (1998), the primary difference being the addition of 18 B-52s and the reduction of B-1s from 50 to 28. Some bombers flying in from CONUS begin arriving relatively early on to support "enabling" operations. The bulk of the bombers arrive in time to support the armor column attacks.

---

[22]See Office of the Secretary of the Air Force (1993).

Finally, we postulate that a minimum airborne component of C3ISR assets (AWACS, JSTARS, Rivet Joint, UAVs) has been flown into theater and are available to provide tactical warning, situational awareness, C2, and targeting support for bomber and fighter forces when combat operations begin.

The allied air forces are assumed to deploy a total of 222 fighters. We postulate that these aircraft were contributed by the U.K., France, Italy, and Germany. All of these nations, except Germany, contributed aircraft in Operations Desert Shield and Desert Storm, and each has compelling economic interests in the region. Each nation deploys between 15 and 20 percent of its fighter force structure to the region, numbers that are in line with historic experience. In future coalition operations, the actual contribution may vary widely.

We did not include fighter forces from the other NATO nations principally because they are unlikely to bring any significant precision interdiction capability against armored vehicles and are unlikely to make significant investments in such weapons by 2010 (see Table 10.2).[23] Although forces from these other nations could support DCA and strike operations, we chose not to include them in the initial phases of the campaign because of the nature of the scenario and the posture of those air forces.

We envision a short-warning scenario that quickly leads to open hostilities and necessitates the rapid deployment of additional fighter forces to the theater. Under these postulated conditions, many of the steps needed to coordinate combat operations of different air forces (establishing command and logistical support relationships, integrating the aircraft into the ATO, etc.) would have to be compressed into a very short time frame.

Given the extraordinary demands on the Joint Force Air Component Command (JFACC), potential theater aircraft beddown constraints, and the limited capabilities offered by the fighters of some of these nations, we assume that the ComCJTF, based on recommendation of the JFACC, limits the number of air forces that participate early on

---

[23]The table indicates that the Netherlands is acquiring advanced air-to-ground precision munitions and may choose to participate in future coalitions. However, its contributions were not considered in this analysis.

rather than accepting all possible contributions. In addition to their interdiction capabilities, the nations we selected to participate in the initial phases of the halt campaign appear to be developing the capability to deploy forces more rapidly than in the past to a greater degree than the remaining countries.

We also postulate that fighter aircraft contributed by the non-NATO allies in the region of conflict are not available for the antiarmor mission in the first couple of weeks of the campaign because they are tied up in DCA. Thus, they are not included in the interdiction analysis.

**Notional Deployment.** Based on the power projection capabilities that the United States has developed, it is normal practice for the United States to prepare for deployments with respect to specific events and decision points. Typically, indications and warning (I&W) trigger the decision process to deploy forces to theater, with the specific deployment based on a time-phased force deployment document (TPFDD); the actual deployment decision day is normally designated as C day. This decision follows substantial deliberations, often time-consuming, among the senior leadership in the U.S. government, often in consultation with its allies, and is made by the national command authority.

U.S. allies undertake similar deliberations and may reach a decision before or after the United States They may decide not to deploy forces early in the conflict or not to deploy forces at all. Decision-making within the NATO Alliance framework is likely to be more complex and subsequently require more time. Another important event in the deployment process is the actual start of hostilities, which is normally designated as D day.

From crisis to crisis, C and D days will vary. By definition, in a short-warning crisis, D day occurs quickly after I&W, and C day hopefully occurs before D day. For this analysis, we envision a relatively short-warning scenario (on the order of several days) that allows for the deployment of U.S. C3ISR assets and a slice of essential support (e.g., tankers, ground control elements, weapons of choice) before hostilities begin. However, the decision to deploy additional U.S. fighter forces to the theater does not occur until D day. Decision times for coalition partners to deploy will vary as well from crisis to crisis.

Because of this wide variability in coalition decision times and to illuminate the importance of strategic interoperability, we postulate U.S. fighter deployments under two distinct conditions. The first condition is when there is support from the NATO allies (ranging from access to bases and overflight to deployments of fighter aircraft). The second is when this support is denied (or the United States is unwilling to wait because the decision will not be timely even if favorable), thereby requiring the deployment of CONUS fighter forces with en route tanker support staged from the United States and at the other end in Saudi Arabia.[24] The latter condition specifically illustrates the importance of strategic interoperability.

With NATO allies' support, the United States can potentially have some CONUS-based fighter squadrons in theater conducting air superiority and interdiction operations within three days of being directed to deploy (in the postulated scenario, that is within three days after hostilities begin). We further estimate that completing the deployment of the fighter forces postulated in this analysis would require about eight days (Appendix D provides more details of our deployment analysis).

This quick-response, out-of-area deployment estimate assumes that the fighters are routed from CONUS bases via great-circle routes over Canada and England into notional German air bases for an en route stopover, then across eastern France into the Mediterranean, and their over Egypt, the Red Sea, and Saudi Arabia to the Doha, Qatar, area in the Middle East—routes that avoid passing over Switzerland, Austria, Eastern Europe, and sensitive Middle Eastern nations. The analysis postulates aerial refueling of fighters by U.S. tankers, staged as needed at NATO allies' air bases.

The analysis also takes into account the refueling of airlifters bringing a notional slice of cargo, with the remaining fighter support either prepositioned or delivered by lift forces prior to the start of combat.

For the second condition, in which allied support (basing, overflight, etc.) is not available, U.S. deployment rates are necessarily delayed.

---

[24]We chose the nonsupported case to determine if it is plausible, what resources are essential to undertake such deployment, and a first-order time line for accomplishing the missions.

While one squadron from CONUS may be available in theater within three days, we estimate that the lack of access to allied support will necessarily delay the completion of the notional U.S. deployment to nearly 14 days. In this case, the fighters fly a great-circle route over the Atlantic through the Strait of Gibraltar, over the Mediterranean, and over Egypt, the Red Sea, and Saudi Arabia to air bases in the Doha area.[25]

Among important factors that can delay the overall deployments are increased operational risk (much longer flight times because of no en route stop), lesser availability of divert air fields, and increased difficulty in coordinating refuelings when tankers from CONUS accompanying the fighters have to be refueled sufficiently to allow for refueling of fighters. For certain fighters, tankers staged in SWA may also be needed. Also, potential diplomatic clearance problems and beddown issues (number of air facilities and associated infrastructure support that the host country is willing to provide at destinations) may arise during deployments. To capture these factors, we assume that only one squadron of fighters can arrive at the destination per day.

U.S. long-range bombers are a unique asset within the NATO alliance—our NATO allies do not have similar capabilities—and provide a significant interdiction capability.[26] The bombers are assumed to initially deploy from CONUS and then return to in-theater locations (e.g., Diego Garcia) after each sortie.

U.S. Naval and Marine Corps F/A-18s deploy from carriers, with the first carrier assumed to be in the theater at the start of combat. An additional carrier arrives seven days later. We assume that one-half of the F/A-18 force is available during the first critical seven days to participate in the halt campaign against the advancing enemy armor columns.

---

[25]Although not considered here, the use of Moroccan airfields as en route stopovers for fighters and airlifters and as tanker operating bases could be an alternative to NATO bases used in the supported case. However, bilateral agreements between the United States and Morocco are required far in advance of hostilities to develop the necessary infrastructure.

[26]Bombers, with their inherent long-range capabilities and large weapon payloads, typically account for a large fraction of the enemy kills in the early phase of a halt operation.

We also considered the contribution of NATO allies' fighter forces in this notional, out-of-area conflict. We assume that their deployment rates are similar to those of the United States, with any shortcomings in their power projection capabilities being offset by their closer proximity to the theater and the availability of air bases en route for refueling and cargo staging. Although deploying at the same rate as U.S. forces would be difficult for the NATO allies today, we assume that they develop the necessary force posture to rapidly deploy selected units in support of this category of operations.[27]

**Force Allocation for Antiarmor.** Our allocation of air assets for this analysis is summarized in Table 11.1, specifically highlighting the contribution of air forces to the antiarmor portion of a notional halt campaign requiring quick response from the United States and its coalition partners. When the antiarmor allocation is combined with the two deployment schedules (i.e., nonsupported and supported cases), we can depict the number of aircraft that are available for the antiarmor mission as a function of time (see Figure 11.4). The value of NATO support is evident. Not only are the CONUS-based fighters deployed to theater faster because of access to allied bases and overflight rights, but the number of aircraft available early in the conflict is even larger when additional support in the form of allied fighters is provided. U.S. bombers used in the antiarmor mission are depicted separately; they are not affected by the level of NATO support. The differences between the two sides of the figure will be major drivers in the results of our analysis.

---

[27]The NATO allies' capabilities for timely engagement in out-of-area operations are limited (see the International Institute for Strategic Studies' *Strategic Balance*). They are generally well endowed with tactical combat (and to a lesser extent with tactical reconnaissance) aircraft. Nevertheless, they lack long-range bombers, and their airlift capabilities are limited in their actual capacity. Coupled with the limited investment in aerial refueling, most NATO allies have limited capability to deploy combat forces to non-European theaters.

Some of the larger NATO nations have noted these shortfalls in recent internal defense reviews and, collectively, the NATO nations (in their Defence Capabilities Initiative) have recognized the need to enhance their ability to rapidly deploy their forces over the next several years. These steps include procuring additional airlifters and fast sealift ships and making changes to the configuration and posture of current units to make them more rapidly deployable.

**Table 11.1**

**Force Allocation**

| Country | Platform | Forces Deployed | Antiarmor Mission | Other Missions[a] |
|---|---|---|---|---|
| United States | F-15E | 66 | 66 | — |
| | F-16CG | 102 | 102 | — |
| | B-1, B-52 | 46 | 38 | 8 |
| | F/A-18C/D | 48 | — | 48 |
| | F/A-18E/F | 48 | 48 | — |
| | F-16CJ | 36 | — | 36 |
| | B-2, F-117 | 26 | — | 26 |
| | F-15C, F-22 | 108 | — | 108 |
| Allies | Tornado IDS | 61 | 31 | 30 |
| | Mirage D/N, 5 | 16 | 8 | 8 |
| | Rafale | 48 | 24 | 24 |
| | EF-2000 | 48 | 24 | 24 |
| | Tornado ECR | 49 | — | 49 |

[a]For example, reactive and lethal SEAD, other strike, and air-to-air missions.

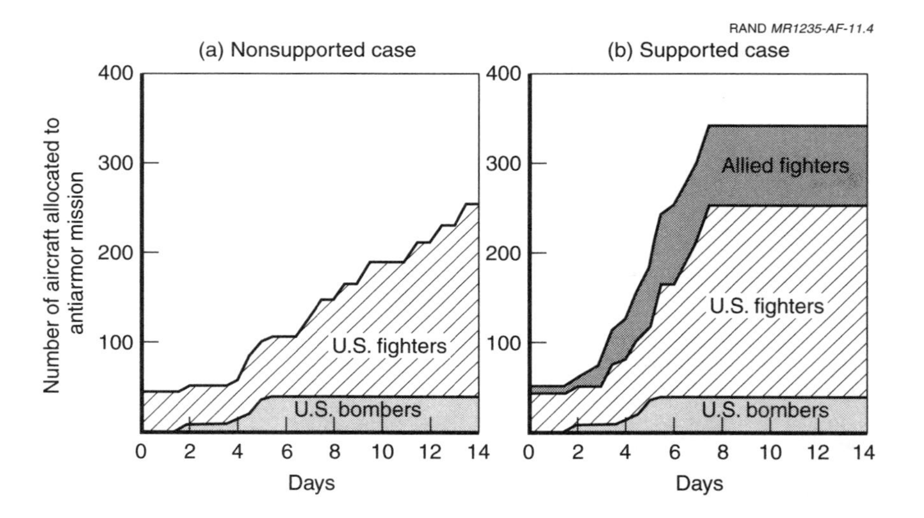


**Figure 11.4—Number of Aircraft Allocated to Antiarmor Mission as a Function of Time**

Sortie rates and weapon carriage are also important inputs, and those used in this analysis are consistent with estimates taken from joint-service studies, other DoD analyses that examined campaign-level effectiveness, and Ochmanek et al. (1998).

As discussed in the preceding section on engagement-level considerations, we postulate that the weapons of choice for the U.S. forces for the halt operation are JSOW and WCMD. JSOW/BLU-108s (the appropriate variant for antiarmor employment) are expected to number 2500 for the Air Force and 900 for the Navy in the 2010 time frame.[28] Additionally, the Air Force is expected to have another 5000 WCMD/BLU-108s, also appropriate against armor targets. Both services should have large quantities of the CEB variant of JSOW, while the Air Force should have large quantities of the CEB variant of WCMD, to attack the nonarmor elements interspersed along the advancing armor columns.

Because the target set contains 50 percent armor targets and 50 percent nonarmor targets, we allocated roughly equal quantities of BLU-108 and CEB variants during each day of the halt campaign. JSOW and WCMD effectiveness were based on the values shown in Figure 11.3. We used the "on-road" values because the armor columns were assumed to be on a road march and the operators on JSTARS were assumed to have accurate road maps. Also implicit in these effectiveness values is the assumption that JSTARS and other ISR assets provide good target discrimination and bomb damage assessment (BDA) to the fighters and bombers.

We assume that the F/A-18s switch to other missions after the Navy inventory of JSOW/BLU-108s is depleted. Similarly, when the Air Force depletes its inventories of JSOW/BLU-108s and WCMD/BLU-108s, we assume that the fighters transition to direct attack PGWs (e.g., Mavericks) and the bombers switch to other missions.

A few U.S. major European allies will have some standoff precision-weapon capability by the year 2010, but most of these weapons will have unitary warheads and thus are unlikely to be used against

[28]Because U.S. military strategy stipulates the ability to conduct two nearly simultaneous MTWs, the United States may choose not to employ all these weapons in a single MTW, as assumed in our analysis.

moving vehicles. Thus, the coalition partners are limited to employing direct-attack PGWs (e.g., Mavericks) against moving vehicles. We assume that they will have less than 1000 during this time frame. The allied fighters conduct other missions once these stocks are depleted. Because these capabilities are limited relative to U.S. capabilities, we considered an excursion in which the NATO allies acquire advanced weapons with antiarmor submunitions such as JSOW and WCMD or their equivalent. In this case, we assume that allied inventories are proportional (relative to force structure size) to those of the United States

**Attrition and Weapon Employment Strategy.** Losses for conventional aircraft against the expected short- and medium-range air defense threats are notionally categorized by weapon delivery profile. The loss rates we used were based on recent operational experience and are consistent with those used in prior RAND studies. For this analysis, we assume that initial (day 1) aircraft losses per sortie would be 0, 0.2, and 2 percent *if* standoff weapon delivery (JSOW), high-altitude direct attack (e.g., WCMD), and low-altitude direct attack (e.g., Maverick) are used, respectively. These loss rates would decrease over time as the air defense capabilities are degraded (e.g., lethal SEAD operations, reactive SEAD such as HARM engagements, and destruction of air defenses interspersed within the column of targeted armored vehicles).

Because aircraft attrition is an overriding concern for all air forces, we assume that the U.S. weapon employment strategy gives preference to standoff weapons such as JSOW over high-altitude direct-attack weapons like WCMD. The strategy is assumed to prefer both of those weapon types over low-altitude direct-attack weapons like Maverick. Using this strategy, U.S. JSOW inventories are sufficient for approximately seven days of the halt campaign (ten days for the no-access case). Thus, no U.S. losses are expected during this first week of JSOW employment. At this point, the Navy fighters switch to other missions and the Air Force fighters and bombers begin using WCMD. Although U.S. losses begin to accumulate during the second week of operations, they remain very low (less than one aircraft after 14 days) because of the success of SEAD operations and the strikes against the armor columns, which include targeting air defenses interspersed along the columns.

The NATO allies, on the other hand, have low-altitude direct attack weapons (e.g., Maverick) and would likely incur significant aircraft losses on a daily basis, perhaps as high as an aircraft per day on average, if they were to use such weapons early in the halt campaign (option one). Although this may be judged acceptable under some circumstances, it is likely that a great many conflicts will not warrant such risks to allied air forces. One alternative (option two) is to delay allied participation in the antiarmor mission until later in the campaign, when the threat from air defenses has been greatly reduced; in this case, their assets would initially be redirected to other operations such as air superiority. Another option (option three) is to assume that the allies have acquired sufficient inventories of JSOW and WCMD or their equivalents so that they may participate early and effectively in the antiarmor mission with risks comparable to those of the U.S. air forces. We considered all three options of allied contributions to the halt campaign but focused on the third option.

**Results.** The value of the interoperability of U.S. and NATO allies' air forces, in the context of our notional multiday air operation against advancing columns of enemy armor, is illustrated in Figure 11.5.[29] We use the term "notional" to indicate that, while there was an attempt to illustrate the results of interdiction missions in the context of an air campaign, the results are not definitive in an absolute sense but are representative in a relative sense and can thus be used to illustrate the military impact of a diverse set of interoperability issues.

Figure 11.5 compares cumulative numbers of enemy AFVs stopped as a function of days for three cases in an out-of-area, quick-response scenario spanning different levels of U.S. and allied participation. As stated earlier, we use as our notional measure of effectiveness the number of days required to stop 50 percent (i.e., 1500) of the enemy's AFVs. In our model, the amount of time needed to achieve this objective is a function of the rates at which JSOW or WCMD carrying BLU-108 submunitions are being delivered. The

---

[29]The results shown in the figure were calculated using a desktop computer model that RAND developed to analyze the halt phase of a campaign. The model was programmed in Analytica™, a visual-modeling system that runs on either a PC or a Macintosh computer. Analytica™ was originally a product of Carnegie-Mellon University and is now distributed, maintained, and extended by Lumina Decision Systems, Inc.

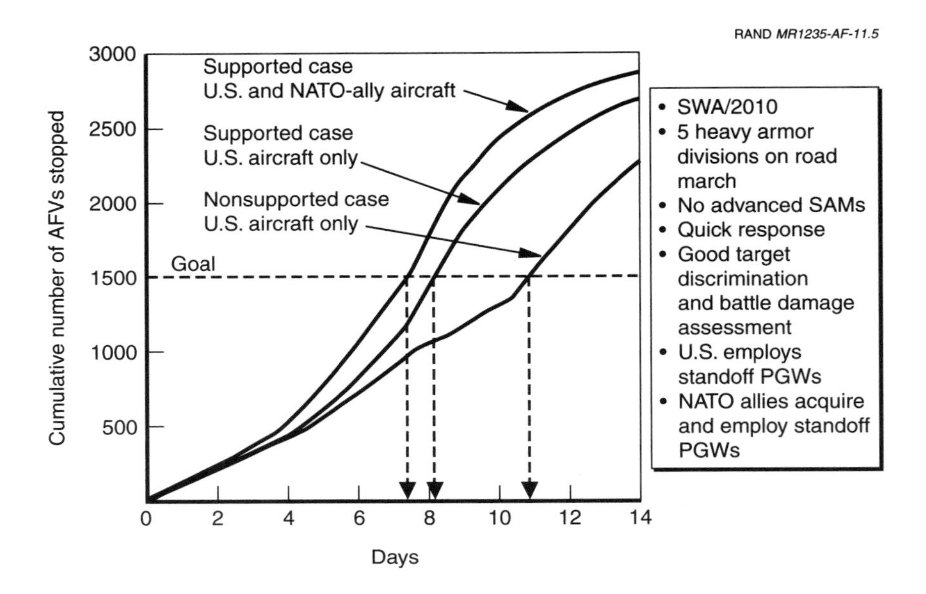


**Figure 11.5—Access to Allied Bases and Airspace May Be More Important than Allied Aircraft/Weapon Contributions**

weapon delivery rates are, in turn, dependent on the sorties rates and weapon carriage for each aircraft type and on the number of aircraft available to conduct the antiarmor mission (from Figure 11.4).

Note that the assessment does not account for enemy armored vehicles that might be damaged or destroyed by regional ground forces or any U.S. predeployed ground forces. It is assumed that the number of such kills will be modest in this type of short-warning scenario. Hence, these results are somewhat conservative.

Also note that these results illustrate halt potential against an enemy defended by the air defense threats described earlier, except that there are no SA-12s. Although we did not explicitly analyze such a scenario, the impact of the SA-12s' presence is discussed later in notional terms.

We first considered a case in which the NATO allies did not contribute forces or allow access to NATO allies' infrastructure or airspace. In this case, the halt campaign is conducted solely by U.S. fighters and bombers. The results illustrate the importance of NATO allies providing access to bases and airspace during the critical first few days of deployment. Without access, it may take nearly 11 days to halt the enemy's advance. With access to NATO allies' infrastructure (the second case), U.S. forces can be deployed more rapidly. As a consequence, the time required to accomplish the halt is reduced by three days. Each such day is critical not only because of the need to limit enemy incursion into friendly territory but also because air assets are in limited supply in the early stages of a conflict and are often needed to respond to other pressing and competing areas of operation simultaneously.

As noted earlier, half the interdiction sorties carried BLU-108 variants of JSOW and WCMD and half carried CEB variants. As an excursion (not shown in Figure 11.5), we also considered an alternative munitions allocation in which 100 percent of the force employs the BLU-108 variants of JSOW and WCMD, switching to the CEB variants only after the stock of BLU-108 variants has been depleted. In this excursion, the halt time decreased by approximately two days with NATO support and three days without NATO support. Even with this new munition allocation, access to NATO allies' infrastructure reduces the time to halt the armor columns by about two days.

The remaining curve in Figure 11.5 (case three) explores the potential impact of additional allied support. In this case we assumed that the NATO allies have acquired advanced weapons with antiarmor submunitions such as JSOW and WCMD or their equivalent (this is the third of three options for allied fighter participation in the halt operation that we examined). We also assumed that allied inventories of JSOW and WCMD are proportional (relative to force structure size) to those of the United States. Because the weapons can be delivered at ranges beyond most of the terminal defenses, their aircraft attrition is low (and comparable to U.S. aircraft attrition while delivering JSOW and WCMD). Allied contributions under this set of circumstances are estimated to shorten the halt operation by a full day. If the NATO allies improve their readiness posture and are able to deploy more forces to theater during the critical halt phase, their contributions would have a larger effect. Results for the other two options for allied fighter participation are not shown because they were less effective than the third option.

The presence of the SA-12 would invariably alter U.S. and allied force allocation strategies and tactics, placing more emphasis early in the campaign on SEAD and in particular on the destruction of the important components of this system (e.g., the radar tracking elements). Until the SA-12 threat is eliminated (which may take several days), armored vehicle attacks by conventional aircraft (U.S. and allied) may be limited owing to their extreme vulnerability to the threat. Since the bulk of these aircraft do not arrive in theater for several days, short delays in SA-12 suppression may not have a significant impact on the time required to halt the armored vehicle column. Longer delays, however, could be more troublesome. In that case, one potential force allocation would be to employ the B-1s and some of the B-2s, delivering JSOW, in order to initiate attacks against the armor columns. Of course, any reallocation of B-2s would probably impact other critical mission areas, namely strategic attack and SEAD.

In summary, the acquisition of advanced weapons would clearly improve overall allied effectiveness in an air campaign against advancing armor columns. However, a substantial allied investment would be required. The level of investment needed is in excess of what NATO allies have historically spent on air-delivered munitions. A more critical benefit that the allies can provide to the United States in conducting such an operation is immediate access to NATO allies' infrastructure and airspace (e.g., tanker basing, overflight rights).

The preceding discussion focused on the halting of enemy armor columns, which is just one aspect of the air campaign. Rather than employ allied air forces to attack moving armor, they could be allocated to target sets better suited to their current air-to-ground capabilities. In particular, we considered allocating their assets to another important mission, the destruction of critical infrastructure and other (mostly) fixed targets that contribute to the enemy's ability to prosecute the invasion. These targets are large in number and diverse in characteristics. The list includes petroleum refineries and pumping stations, aircraft production facilities, power plants, bridges, railroad yards and their associated facilities. It also includes important communications and C2 nodes.

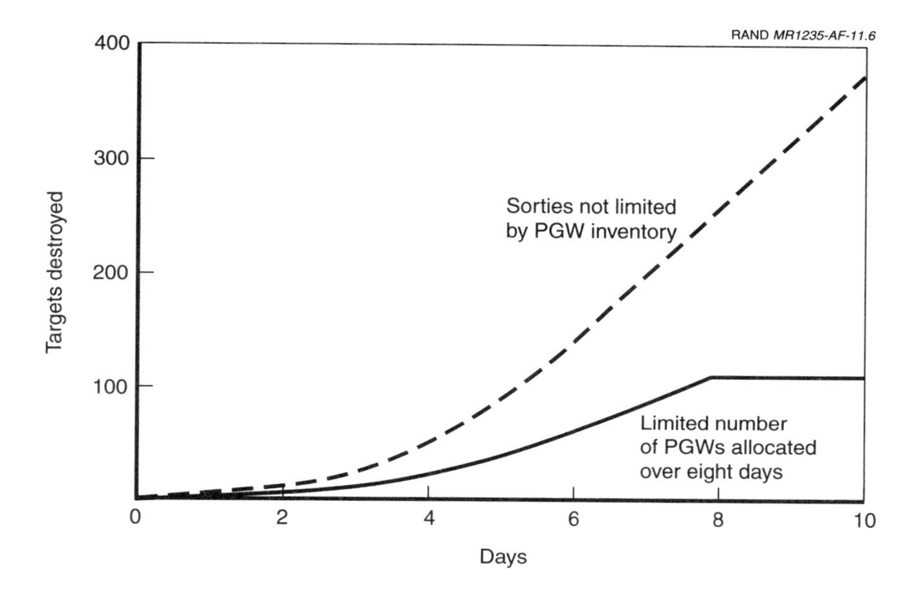


**Figure 11.6—Critical Infrastructure Targets Destroyed by Allies**

Figure 11.6 shows the potential numbers of these targets destroyed by allied air forces using PGWs with unitary warheads, assuming the NATO allies' air forces that were once allocated to attacking moving armor are now reallocated to critical infrastructure targets. If these targets are defended by highly capable air defenses, substantial SEAD assets would be required to conduct such missions. If the sorties are not limited by PGW inventories (assuming that the allies purchase additional laser-guided bombs or GPS-guided munitions like JDAM), they have the potential to destroy about 350 of these targets over a ten-day period.[30] In cases where their inventories are limited, the estimated number of targets destroyed will be less. In this latter example, we assume a total inventory of 1000 allied PGWs delivered over an eight-day period, matching the approximate length of the concurrent halt campaign.

---

[30]The number of weapons needed to destroy the wide range of targets varies greatly depending on the number of aim points and the weapon effectiveness.

## OBSERVATIONS

U.S. and its NATO allies' air forces are and likely will continue to be adequately interoperable in air-to-air operations, particularly in air surveillance, air combat patrol, no-fly-zone enforcement, and DCA missions against aircraft and moderately stealthy cruise missiles. Interoperability should improve over time with the integration of MIDS and upgrades to the AWACS fleets. However, the United States is well ahead of its allies in developing ballistic missile defense capabilities and therefore currently bears a major burden in force protection of coalition forces. To minimize the risk of allied forces to ballistic missiles attacks, the allies should leverage U.S. investments in this area and pursue complementary efforts such as interoperable communications and data exchange systems, interoperable radars, and weapon systems.

The increasing gap between the United States and its allies in all-weather, standoff PGWs and smart submunitions for attacks on moving targets poses a greater interoperability challenge in future coalition operations. If U.S. allies do not commit to proportionately comparable investments in such capabilities, the role of their air forces in future air-to-ground operations, which increasingly demand attainment of military objectives with minimal casualties and collateral damage, will likely decrease substantially.

Currently, the NATO allies' air forces are not configured to rapidly deploy a substantial number of fighters to out-of-NATO-area operations.[31] Improvements in force posture, airlift, and aerial refueling capabilities need to be made to support a strategy that includes quick intervention in conflicts.

The United States greatly benefits from allies' support for deployment operations such as the one envisioned in this campaign.

---

[31]In Operation Allied Force, NATO employed large numbers of fighter aircraft in a contingency that was technically out of area. However, many of the aircraft were able to operate from their home bases because of the bases' proximities to the area of operations. Of the aircraft that did deploy, all were able to use existing NATO air bases. In addition, NATO had been operating in the area for several years and was given ample warning. These circumstances may not be repeated in future out-of-area operations that are some distance from NATO territory, such as possible future air operations in SWA.

Access to allies' airspace, air bases, and infrastructure is crucial to such deployments. These strategic interoperability benefits are vital to future U.S. interests and should not be jeopardized by operational and tactical interoperability issues.

Chapter Twelve

# CONCLUDING OBSERVATIONS AND SUGGESTED ACTIONS

Interoperability not only supports U.S. national security and U.S. military strategies but also fosters and enables allied support for coalition operations. It can increase the effectiveness and efficiency of U.S. and allied forces in such operations.

The United States and its partners have many shared interests and concerns. Thus, it is important that these countries' forces are able to operate effectively and efficiently in coalitions to achieve common goals. This is especially critical in air power, where the speed of operations dictates close harmonization of effort to achieve maximum impact on the battle outcome. Interoperability at the technological, tactical, and operational levels is key to achieving this close harmonization. In some cases, coalition support (e.g., access to allied air space, en route bases, and infrastructure support) is required for the United States to conduct successful military operations. In such cases, interoperability at the strategic level is required.

## RECENT OPERATIONS

Our review of recent coalition operations indicates that interoperability has multiple and complex dimensions—political and economic as well as military—that may manifest themselves at strategic, operational, tactical, and technological levels. Further, the impact of interoperability problems is not isolated within the level at which they are observed. Strategic-level interoperability problems can have operational and tactical implications, and technological interoperability problems may reverberate in the opposite direction. For example, the political and economic goals of individual nations to support national industries can lead to development of air power systems (e.g., fighters, weapons, airborne surveillance and control assets) that have different capabilities and require extensive workarounds to be employed in coalition operations. Similarly, the lack of interoperable communications and combat identification systems and procedures could result in the attrition of coalition aircraft to enemy defenses or, through unfortunate Blue-on-Blue engagements, cause the partner to leave the coalition. This suggests that interoperability issues should be considered in the context of each level.

Political support, access to allied infrastructures and airspace, landing rights, and forward basing are essential to bringing U.S. air power to bear effectively in certain regions of interest. Specifically, allied support for and participation in coalitions help U.S. decisionmakers garner and maintain the public support necessary to conduct military operations in regions of the world that are of national interest (e.g., SWA and the Balkans). Moreover, as seen in recent Allied Force Balkan operations, access to allied airspace and availability of infrastructure in close proximity to areas of operations minimized flight time to air patrol stations and targets and provided flexibility to conduct attack operations from more than one approach azimuth.

The factors above are sometimes overlooked when potential contributions of individual allies are measured solely in terms of the number of aircraft made available or sorties flown in specific operations. While it is true that in recent SWA operations (e.g., Desert Storm, Desert Fox) the United States provided the vast majority of air missions, in some Balkan coalition operations—particularly those not involving precision strike operations (e.g., air surveillance, airspace control, and no-fly-zone enforcement)—allies provided more than half of the aircraft and missions flown. Further, it is important to recognize that providing the preponderance of air assets to coalition operations helped the United States rationalize its smaller ground force contributions in Bosnia and Kosovo operations.

The above notwithstanding, allied contributions to recent strike operations in the Balkans have been limited because of the lack of sufficient PGWs that can be delivered day or night in any weather conditions. A number of U.S. allies are developing plans to expand their holdings of PGWs, including those that are guided by GPS. It is critical that these be implemented, since future crises can be expected to require the use of PGWs to minimize casualties and collateral damage and the use of standoff weapons to minimize the risk of attrition of coalition aircraft to more sophisticated enemy air defenses.

Another future concern is the allies' limited capabilities (e.g., force readiness, airlift) to rapidly deploy forces to out-of-area operations. The allies have made great strides in their ability to deploy and support operations outside their borders to the periphery of Europe, but more improvement is needed if they must rapidly deploy combat forces to non-European theaters. Thus, without improvements to existing capabilities, the combat value of allied air forces is likely to decrease in the future.

## CASE STUDIES

The case studies examined suggest that the following areas offer the best leverage for achieving acceptable levels of interoperability in future coalition operations: (1) common or harmonized doctrine for the planning, execution, and execution monitoring of CJTF operations in general and air campaigns in particular; (2) compatible or adaptable concepts of operations and procedures for airborne surveillance and control in support of air-to-air and air-to-ground missions; (3) common information-sharing standards and compatible tactical communication systems; and (4) expert personnel who understand the capabilities of coalition partners and who hone their expertise in combined operations and exercises.

Efforts to enhance interoperability solely through common or fully interoperable systems at the technological level are likely to be limited by political, economic, and security factors. Perhaps the most important factors are support of national industries, equitable burden sharing, and ensuring that the most advanced military capabilities are not compromised. From a technology and cost perspective, selected C3ISR initiatives appear to offer the best opportunities for interoperability enhancements.

## SUGGESTED ACTIONS

Our review of recent coalition operations, the results of case studies, and the survey of allies' capabilities suggest a range of actions to foster interoperability and minimize divergence between U. S. and allied air forces. Because of the broad nature of interoperability and its implications to a wide range of stakeholders (e.g., other U.S. services and NATO partners), the Air Force has to address many interoperability issues in collaboration with such stakeholders. Some actions the Air Force can undertake directly include the following.

### Collaborative Actions

In collaboration with DoD, other U.S. services, and NATO allies, the U.S. Air Force should

- Help NATO develop the CJTF CONOPS, associated processes, expert personnel, systems, and information-sharing protocols for out-of-area operations. In particular, the Air Force should ensure that the key doctrinal concept of centralized control and decentralized execution, which is inherent in U.S. joint-service air CONOPS, is institutionalized in the NATO CJTF concept.
- Help NATO define the desired level of information sharing and interoperability between planned U.S. and NATO force-level planning and execution-monitoring capabilities (organizations, procedures, personnel, and systems). At a minimum, a set of common messaging standards for information exchange should be defined for the U.S. Air Force's TBMCS and NATO's ICC and for the TBMCS and NATO's ACCS.
- Help NATO develop a coherent space policy and information sharing protocols that provide sufficient information to conduct key operations, while protecting sensitive equities. In some cases, bilateral agreements with selected NATO allies may be more appropriate.
- Continue to foster the interoperability of AWACS assets and standard AEW and control procedures, especially those needed in the presence of friendly and enemy stealth aircraft. The focus should be on ensuring that NATO, U.K., French, and U.S. AWACS modernization programs are synchronized.

- Develop the process and capabilities to receive and exploit ground surveillance information from the different airborne GMTI sensors that NATO members are developing. Support advanced concept demonstrations to determine the value of this capability and help select the most appropriate means to achieve it, including development of common GMTI data formats.[1]
- Ensure that the MIDS EMD program is successfully completed and that the functional interoperability inherent in MIDS terminals is maintained through the production phase and then applied to future fighter data links.
- Continue to share fighter and weapon systems information to ensure adequate common understanding of individual coalition partners' air capabilities (technology, personnel, operations). In parallel, continue to develop operating protocols that permit the use of allied air assets in coalition operations and expand training exercises to emphasize out-of-area operations. Be prepared to employ workarounds.
- Encourage NATO allies' acquisition of advanced precision weapons and standoff weapons. Low-cost GPS-guided weapons are particularly promising. Although they are expensive, standoff weapons ensure platform survivability in a high-threat environment, and standoff antiarmor weapons enable more effective participation in the halt phase of a campaign.
- Increase opportunities for combined experiments and advanced technology demonstrations.
- Support the above suggested actions by actively participating in NATO's Defence Capabilities Initiative.

## Direct Actions

In parallel with the preceding collaborative efforts, the U.S. Air Force should consider taking the following direct actions:

- Leverage its expertise and capabilities in planning and executing air and space operations in power projection missions by man-

---

[1]These are appropriate actions given the uncertainties of the NATO AGS program.

ning key positions in the emerging deployable and key static CAOCs to reinforce the principle of centralized control and decentralized execution. Further, it should develop and maintain a cadre of experts who can provide support to higher NATO headquarters (if needed) to help develop air campaign plans and assist in execution monitoring.

- Explore opportunities to gain better visibility into the WEU Satellite Centre to determine if and how Centre assets might help satisfy some of the information needs in future NATO operations.
- Ensure that the AWACS RSIP program continues to be adequately funded and that appropriate NATO RSIP employment lessons learned are incorporated in future early warning and air control doctrine and tactics.
- Support advanced concept technology demonstration of multiple GMTI sensor data reception and exploitation capabilities in joint expeditionary force experiments.
- Strengthen Air Force visibility and management oversight in the MIDS production phase program to ensure that MIDS terminals are delivered as needed to U.S. fighter modernization programs, within budget constraints.

The collaborative and direct actions identified in this research, if successfully executed, are likely to improve the military performance of U.S. and NATO allies' air forces in future coalition operations or NATO Alliance operations.

Appendix A

# ALLIES' PARTICIPATION IN AND CONTRIBUTIONS TO RECENT COALITION OPERATIONS

We reviewed a number of recent coalition operations (1) to better understand the dimensions, issues, and value of interoperability; (2) to identify the sorts of challenges that can arise in coalition operations and provide a starting point for understanding and addressing interoperability in future coalition operations in the new security environment, in general and in the various case studies in particular; and (3) to lay the groundwork for a discussion of the benefits and costs of coalitions and interoperability.

This appendix presents a short summary of our review of 40 recent coalition operations that included NATO allies and one NATO Alliance operation. It addresses the missions for which interoperability is required, the allies' participation in recent operations, and the contributions that the allies provided.[1]

## MISSIONS FOR WHICH INTEROPERABILITY IS REQUIRED

As suggested by Table A.1, recent history reveals not only that the United States has operated in coalitions across the entire "spectrum of conflict"—from humanitarian relief and peacekeeping operations in a permissive environment to MTW—but that non-MTWs predominate.

[1]For more detailed information on our analysis of recent operations, see Larson et al. (1999).

**Table A.1**

**Forty U.S. Multilateral Operations by Mission Focus**

| Mission Focus | Non-U.N. | U.N. |
|---|---|---|
| Humanitarian | 4 | 1 |
| Peacekeeping | 3 | 11 |
| Monitoring/observation | 5 | 2 |
| Airlift | 2 | |
| No-fly zones | 4 | |
| Other peace enforcement | 1 | |
| Crisis responses | 3 | |
| Strike operations | 3 | |
| Major theater war | 1 | |
| Total | 26 | 14 |

This observation is based on analysis of 14 recent United Nations operations and 26 non-U.N. operations (listed in Tables A.2 and A.3) in which the United States operated in coalitions including Operation Allied Force, the only Alliance operation with NATO partners. This recent historical experience dictates that interoperability issues be considered across the entire spectrum of U.S. military operations, and that robust measures—i.e., those that enhance interoperability across a wide range of missions—will generally be preferred over more tailored solutions.

## NATO ALLY PARTICIPATION IN RECENT COALITIONS

### Providing Forces

Although participation in coalition operations has varied greatly from situation to situation and over time (see Table A.4), a number of allies have been particularly reliable in their participation in recent coalitions in which the United States has also participated.

As shown in Table A.4, the most frequent NATO coalition partners in the 40 operations examined were the United Kingdom (29 of 40 operations), France (28), Turkey (23), Germany (22), and Italy and the Netherlands (21 each). Other NATO allies participated in fewer actions with the United States.

**Table A.2**

**Twenty-Six Recent U.S. Non-U.N. Multilateral Operations**

| Operation Name | Location | Mission | Date |
|---|---|---|---|
| Provide Promise | Fmr. Yugo. | HR | Jul 92–Mar 96 |
| Maritime Monitor | Adriatic | MON | 06.16.92–11.22.92 |
| Sky Monitor | Bosnia | MON | 10.16.92–04.12.93 |
| Deny Flight | Bosnia | NFZ | 04.12.93–12.20.95 |
| Sharp Guard | Adriatic | MON | 06.15.93–10.02.96 |
| Quick Lift | Croatia | LIFT | July 1995 |
| Deliberate Force | Bosnia | STR | 08.29.95–09.21.95 |
| Joint Endeavor (IFOR) | Bosnia | PK/PE | 12.20.95–12.20.96 |
| Decisive Enhancement | Adriatic | MON | Dec 95–06.19.96 |
| Decisive Edge | Bosnia | NFZ | Dec 95–Dec 1996 |
| Determined Guard | Adriatic | PK | Dec 96–present |
| Joint Guard (SFOR I) | Bosnia | PK | 12.20.96-6/20/98 |
| Joint Forge (SFOR II) | Bosnia | PK | 6/20/98–present |
| Determined Force | Kosovo | CR | Planned Sept 1998 |
| Eagle Eye | Kosovo | MON | 10.16.98–present |
| Allied Force | Kosovo | STR | 3/25/99–6/20/99 |
| Desert Storm | SWA | MTW | 01.17.91–02.28.91 |
| Provide Comfort | Kurdistan | HR | 04.05.91–12.31.96 |
| Southern Watch | Iraq | NFZ | Aug 1992–present |
| Vigilant Warrior | Kuwait | CR | Oct 1994–Nov 1994 |
| Northern Watch | Iraq | NFZ | 12.31.96–present |
| Desert Thunder | Iraq | CR | 09.03.96–09.04.96 |
| Desert Fox | Iraq | STR | 12.16.98–12.19.98 |
| Quick Lift | Zaire | LIFT | 09.04.91–Oct 1991 |
| Restore Hope | Somalia | HR | 12.11.92–05.04.93 |
| Guardian Assistance | Zaire/Rwanda | HR | 11.14.96–12.27.96 |

NOTES: HR = humanitarian relief; MON = monitoring/observation; LIFT = airlift; PK = peacekeeping; PE = peace enforcement; NFZ = no-fly zone; CR = crisis response; STR = strike; MTW = major theater war.

The implications are twofold. The first is that interoperability planning must be adaptive enough to accommodate the possibility of coalitions of different sizes and composed of different coalition partners. "Plug-and-play" is a concept that is well known at the technological level. But it is also required at the national level to provide for the possibility of different combinations of coalition partners; to manage the comings and goings of coalition members as the mission focus changes and/or missions are added, completed, or abandoned; and to minimize disruptions to the overall coalition effort. This suggests a focus on long-term interoperability solutions,

**Table A.3**

**Fourteen Recent U.N. Operations with U.S. Participation**

| Operation | Location | Mission | Date |
|---|---|---|---|
| UNPROFOR | Former Yugoslavia | PK | Feb 92–Mar 95 |
| UNCRO | Croatia | PK | Mar 95–Jan 96 |
| UNPREDEP | Macedonia | PK | Mar 95–present |
| UNMIBH | Bosnia | PK | Dec 95–present |
| UNTAES | Croatia | PK | Jan 96–Jan 98 |
| UNPSG | Croatia | PK | Jan 98–present |
| UNTSO | Jerusalem | MON | Jun 48–present |
| UNIKOM | Iraq/Kuwait | MON | Apr 91–present |
| UNAMIC | Cambodia | PK | Nov 91–Mar 92 |
| UNTAC | Cambodia | PK | Mar 92–Sep 93 |
| UNOMIG | Georgia | PK | Aug 93–present |
| MINURSO | Sahara | PK | April 1991–present |
| ONUMOZ | Mozambique | PK | Dec 92–Dec 94 |
| UNOSOM II | Somalia | HR | Mar 93–Mar 95 |

NOTES: PK = peacekeeping; MON = monitoring/operation; HR = humanitarian relief.

including organizations, doctrine, procedures, and system architectures that can accommodate the dynamic character of coalitions, including transitions.

The second implication is that because the United States' NATO allies vary in their coalition participation with the United States, the United States might be able to achieve important interoperability through a series of bilateral rather than alliance-wide efforts.

## Providing Base Access

In addition to providing forces, coalition members can also provide other types of services and resources; from the vantage point of air power, perhaps the most important of these is base access and support. Although there is a great deal of variance in the provision of bases from operation to operation, of particular interest is the consistent support that Germany and Italy have provided in recent operations in the Balkans.

**Table A.4**

**NATO Participation in U.S. Multilateral and U.N. Operations**

| Country | U.N. | Non-U.N. | Total |
|---|---|---|---|
| United States | 14 | 26 | 40 |
| Belgium[a] | 8 | 9 | 17 |
| Canada | 11 | 8 | 19 |
| Czech Republic[b] | 1 | 0 | 1 |
| Denmark[c] | 9 | 5 | 14 |
| France[a] | 10 | 18 | 28 |
| Germany[a] | 7 | 15 | 22 |
| Greece[a] | 5 | 11 | 16 |
| Hungary[b] | 2 | 1 | 3 |
| Iceland[d] | 1 | 0 | 1 |
| Italy[a] | 7 | 14 | 21 |
| Luxembourg[a] | 0 | 1 | 1 |
| Netherlands[a] | 8 | 13 | 21 |
| Norway[d] | 10 | 8 | 18 |
| Poland[b] | 3 | 0 | 3 |
| Portugal[a] | 6 | 8 | 14 |
| Spain[a] | 4 | 11 | 15 |
| Turkey[d] | 7 | 16 | 23 |
| United Kingdom[a] | 7 | 22 | 29 |

NOTES: "U.N." signifies United Nations operations in which the United States participated with other NATO allies; "non-U.N." is non-U.N. U.S. coalitions that included NATO allies.

[a]Also member of Western European Union.

[b]Joined NATO in 1999.

[c]WEU observer.

[d]Associate member of WEU.

## CAPABILITIES CONTRIBUTED TO RECENT COALITIONS

Based on the operations examined, allied contributions appear to vary greatly across operations. As shown in Figure A.1, in SWA, the United States historically has contributed a majority of the aircraft, while in many Balkans operations NATO allies have contributed a majority.

The United States not only is often the single largest contributor to coalition operations but also tends to contribute the broadest range of aircraft (see Figure A.2). Nevertheless, several nations—the United

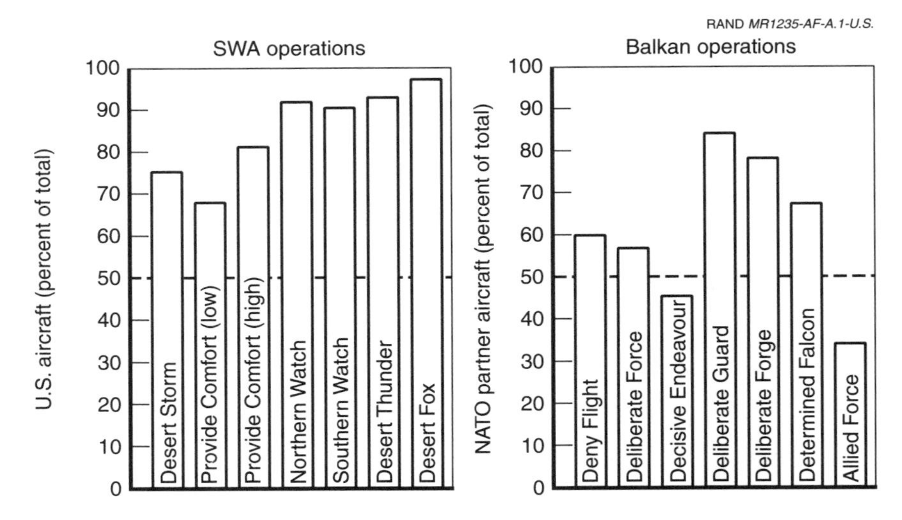


NOTE: See Table A.2 for a listing of the locations and dates of the operations presented in this figure. See Larson et al. (1999) for a more detailed description of these operations.

**Figure A.1—U.S. Aircraft Contributed to SWA and Balkan Operations**

Kingdom, France, and Italy—also have some breadth in their air capabilities.

As shown in Table A.5, which describes the U.S. and coalition sorties flown in Operation Desert Storm, the broad air power capabilities of the United States allow its air forces the flexibility and robustness to fly the widest range of combat missions.

These observations suggest that important roles can be and are being played by the United States' coalition partners, and U.S. interoperability planning should take advantage of these capabilities. Nevertheless, because coalition partners vary across operations, the United States may often need to provide the richest mix of forces—or the C3ISR backbone—so as to provide the "glue" for planning and executing the operation.

The examination of recent coalition operations also reveals that non-weapon-system contributions (e.g., access to and use of forward

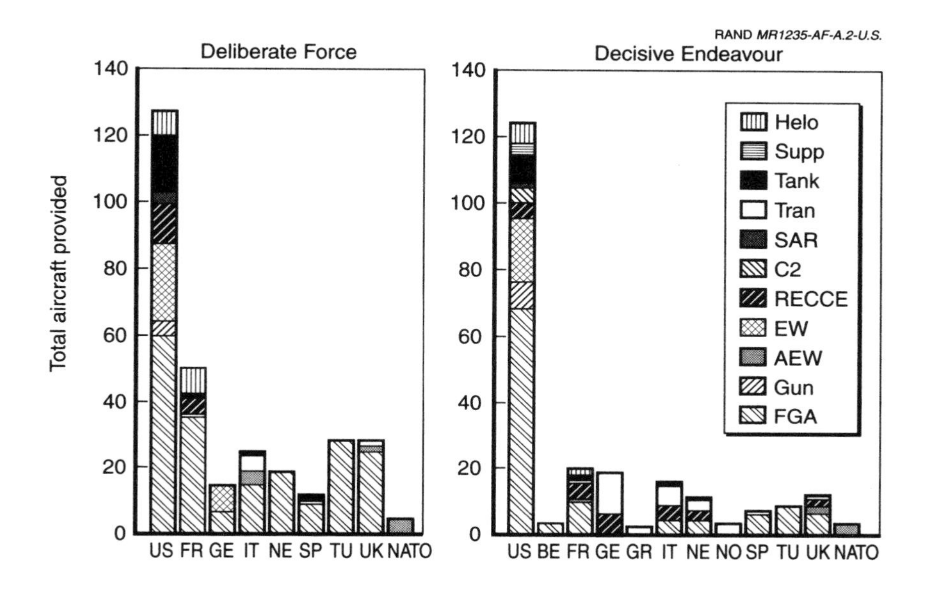


**Figure A.2—The United States Brings a Broader Range of Capabilities**

air bases for beddown and operation of aircraft, infrastructure, tanker support, and airspace) can be critical contributions that can enhance coalition interoperability.

## Table A.5

## U.S. and Coalition Sorties Flown in Operation Desert Storm

| Country | AI | CAS | CAP | SCAP | OCA | C3 | RECCE | EW | SOF | LIFT | TANK | SUPP | TRAIN | Other | Total |
|---|---|---|---|---|---|---|---|---|---|---|---|---|---|---|---|
| U.S. total | 33,648 | 6,128 | 8,803 | 198 | 9,115 | 1,904 | 2,894 | 2,856 | 946 | 17,657 | 14,323 | 1,022 | 526 | 1,368 | 101,388 |
| USAF | 24,292 | 2,120 | 4,558 | 0 | 6,422 | 604 | 1,311 | 1,578 | 134 | 16,628 | 11,024 | 203 | 174 | 358 | 69,406 |
| USN | 5,060 | 21 | 4,245 | 198 | 1,936 | 1,143 | 1,431 | 265 | 3 | 0 | 2,782 | 41 | 262 | 916 | 18,303 |
| USMC | 4,264 | 3,956 | 0 | 0 | 757 | 157 | 3 | 343 | 1 | 9 | 461 | 714 | 14 | 4 | 10,683 |
| USSOCCENT | 32 | 31 | 0 | 0 | 0 | 0 | 2 | 84 | 808 | 19 | 56 | 64 | 76 | 90 | 1,262 |
| USA | 0 | 0 | 0 | 0 | 0 | 0 | 147 | 586 | 0 | 201 | 0 | 0 | 0 | 0 | 934 |
| CRAF | 0 | 0 | 0 | 0 | 0 | 0 | 0 | 0 | 0 | 800 | 0 | 0 | 0 | 0 | 800 |
| NATO allies | 1,970 | 0 | 1,729 | 40 | 1,264 | 0 | 218 | 80 | 1 | 2,529 | 1,087 | 40 | 158 | 98 | 9,214 |
| Canada | 48 | 0 | 693 | 0 | 144 | 0 | 0 | 0 | 0 | 277 | 64 | 0 | 64 | 12 | 1,302 |
| France | 531 | 0 | 340 | 0 | 230 | 0 | 62 | 0 | 1 | 855 | 223 | 0 | 4 | 12 | 2,258 |
| Italy | 135 | 0 | 0 | 0 | 0 | 0 | 0 | 0 | 0 | 13 | 89 | 0 | 0 | 0 | 237 |
| UK | 1,256 | 0 | 696 | 40 | 890 | 0 | 156 | 80 | 0 | 1,384 | 711 | 40 | 90 | 74 | 5,417 |

**Table A.5—continued**

| Country | AI | CAS | CAP | SCAP | OCA | C3 | RECCE | EW | SOF | LIFT | TANK | SUPP | TRAIN | Other | Total |
|---|---|---|---|---|---|---|---|---|---|---|---|---|---|---|---|
| GCC allies | 2,659 | 0 | 2,543 | 0 | 291 | 8 | 124 | 0 | 1 | 1,878 | 485 | 9 | 2 | 0 | 8,000 |
| Saudi | 1,656 | 0 | 2,391 | 0 | 277 | 8 | 118 | 0 | 0 | 1,829 | 485 | 9 | 2 | 0 | 6,775 |
| Kuwait | 780 | 0 | 0 | 0 | 0 | 0 | 0 | 0 | 0 | 0 | 0 | 0 | 0 | 0 | 780 |
| Bahrain | 122 | 0 | 152 | 0 | 14 | 0 | 0 | 0 | 1 | 4 | 0 | 0 | 0 | 0 | 293 |
| UAE | 58 | 0 | 0 | 0 | 0 | 0 | 6 | 0 | 0 | 45 | 0 | 0 | 0 | 0 | 109 |
| Qatar | 43 | 0 | 0 | 0 | 0 | 0 | 0 | 0 | 0 | 0 | 0 | 0 | 0 | 0 | 43 |
| Grand total | 38,277 | 6,128 | 13,075 | 238 | 10,670 | 1,912 | 3,236 | 2,936 | 948 | 22,064 | 15,895 | 1,071 | 686 | 1,466 | 118,602 |

SOURCE: *Gulf War Air Power Survey,* Vol. V, Washington, D.C.: Government Printing Office, 1993, Table 64, "Total Sorties by U.S. Service/Allied Country by Mission Type," pp. 232–233.

NOTE: AI = air interdiction; CAS = close air support; CAP = combat air patrol; SCAP = surface combat air patrol; OCA = offensive counterair; C3 = command, control and communications; RECCE = reconnaissance; EW = electronic warfare; SOF = special operations forces; LIFT = airlift; TANK = aerial refueling; SUPP = support; TRAIN = training.

Appendix B

# NEW OPERATIONAL CONCEPTS FROM *JOINT VISION 2010*

Four new operational concepts are described in *Joint Vision 2010:* precision engagement, dominant maneuver, focused logistics, and full-dimensional protection. The following definitions are taken from the Chairman of the Joint Chiefs of Staff (1996):

*Precision engagement* consists of a system of systems that enables our forces to locate the objective target, provide responsive command and control, generate the desired effect, assess our level of success, and retain the flexibility to reengage with precision when required. Among others, precision engagement envisions (1) a substantial increase in the use of all-weather, precision-guided stand-off weapons as compared to recent operations and (2) rapid combat assessment and restrike of targets if necessary.

*Dominant maneuver* is defined as the multidimensional application of information, engagement, and mobility capabilities to position and employ widely dispersed joint air, land, sea, and space forces to accomplish assigned operational tasks. This concept envisions a nonlinear battlefield where, for example, light ground forces, supported by air, sea, and space force, operate across the breadth and depth of the battlefield—a radical departure from traditional linear battlefield operations.

*Focused logistics* is defined as the fusion of information, logistics, and transportation technologies to provide rapid crisis response, to track and shift assets even while en route, and to deliver tailored logistics packages and sustainment directly at the strategic, operational, and tactical levels of operations.

*Full-dimensional protection* is defined as the control of the battle space to ensure that our forces can maintain freedom of action during deployment, maneuver, and engagement while providing multilayered defenses for our forces and facilities at all levels. Current efforts to build an integrated defense-in-depth theater air and missile defense capability constitute an example of a full-dimensional protection component.

Appendix C

# MIDS CASE STUDY

The Multifunctional Information Distribution System (MIDS) case study is different from the other case studies in which potential solutions to interoperability problems are analyzed and discussed. In this case study, the near-term solution for an interoperable communication system has already been selected, and it is MIDS.[1] Thus, this case study is really an acquisition case study that highlights the programmatic complexities of cooperative initiatives designed to enhance interoperability among coalition forces.

A summary of the case study was provided in Chapter Nine. There we describe the three major reasons for the MIDS program, summarize our observations of the case study, and present suggested actions the Air Force could take to ensure the success of the MIDS pro-

---

[1]One of the drawbacks of MIDS, which is shared by other Link 16 terminals, is an aging system design that takes limited advantage of recent technology developments. This case study does not address the issue of whether this program—or, for that matter, JTIDS—will support all fighter data link needs in future military operations. As discussed in our past work (Hura et al., 1998), additional research on this larger issue is warranted. This case study focuses on short-term solutions to urgent operational requirements. More capable and more technologically advanced data link systems such as the Joint Tactical Radio System (JTRS) are under development by the DoD and may meet the more stressing far-term needs of the services. However, JTRS will not be available in the near term. On the other hand, if the MIDS program can be transitioned into the production phase without major delays, the urgent data link requirements of the MIDS program member nations can be satisfied in the near term.

After this research was completed, additional information regarding enhancements to Link 16 became available. In particular, the U.S. military is investigating enhanced throughput (higher data rates) and dynamic network management for Link 16 (Simkol, 2000). These enhancements would mitigate some of the current shortfalls of Link 16.

gram. Because of the complexity of the MIDS program and because there is a separate report on the case study,[2] most of the details are presented in this appendix. Here we examine the goals of the program and the MIDS terminal architectures; discuss programmatic issues, including the history of the program over the last decade; review how MIDS grew out of the original U.S. Air Force–led JTIDS joint-service program; discuss projected costs of MIDS production terminals; and compare those costs to the possible costs of JTIDS Class 2R production terminals if the latter program had proceeded as originally envisioned by the Air Force.

## MIDS PROGRAM GOALS AND TERMINAL ARCHITECTURE

### Goals

The first goal of the MIDS program was to develop a modular open terminal architecture. With an open architecture it will be easier to integrate MIDS terminals into dissimilar platforms built by different contractors.[3]

The second goal was to develop an affordable terminal that could be readily tailored to fit any military platform. Initially, MIDS terminals were developed for integration into a set of platforms specified by participating member nations.[4] Later, the MIDS architecture was modified to accommodate additional U.S. aircraft.

The final and operationally most significant goal of the program was to provide interoperable, jam-resistant[5] C2 data communication links between U.S. and allied platforms, regardless of whether they

---

[2]See Gonzales et al. (2000).

[3]This "open" architecture predates the Defense Information Infrastructure (DII) common operating environment (COE) and is distinct from it in that it is not specifically based on a set of open software or hardware standards. However, it is partially based on commercial standards for real-time processing systems and on the use of widely available commercial components, such as the Motorola 68040 microprocessor.

[4]In the early days of the MIDS program, the smallest aircraft in the inventory of MIDS member nations was the F-16. Compatibility with aircraft like the F-16 has been a driving program requirement.

[5]In this study, we do not address what level of jam 200resistance is likely to be required in future environments.

were ground-based C2 nodes, ships, or fighter/bomber aircraft. Such interoperability would be ensured by MIDS because participating member nations would be required to acquire MIDS terminals for their military forces.

## LVT Terminal Architecture

Shown in Figure C.1 are key elements of the MIDS architecture for the LVT, the original terminal of the MIDS program. The LVT-platform integration approach is illustrated in the left-hand figure. The LVT is connected to the platform avionics bus (e.g., the 1553 bus of an F-16). Through the bus it exchanges information with platform systems, including cockpit input/output (I/O) devices such as numeric keypads, cockpit displays showing air threats or targets, communications and navigation antennas, and onboard processors. The MIDS architecture enables the LVT to exchange information with such systems on the specified platforms of participating member nations.

The MIDS LVT hardware architecture is illustrated by the middle diagram in the figure. The basis of this architecture is the LVT chassis, which is common to all MIDS platforms. The chassis holds up to nine standardized electronic cards, or Standard Electronic Modules Format-E (SEM-E), each with specific functionality such as voice or message-processing functions. The cards can easily be replaced in

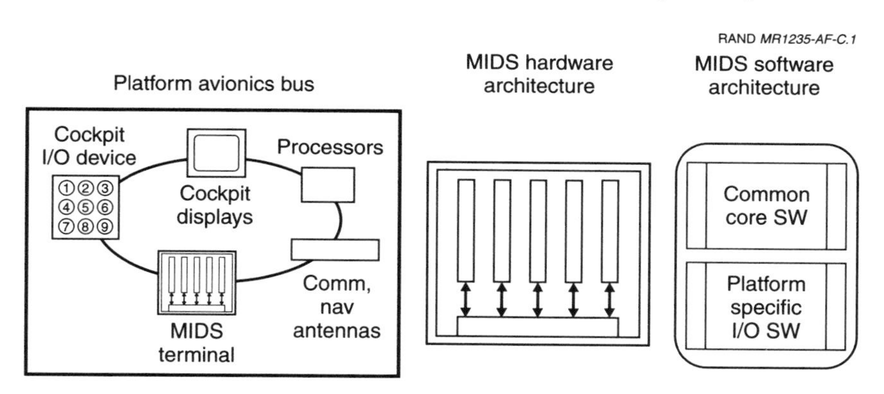


**Figure C.1—MIDS LVT Terminal Architecture**

the event of failure or if a specific functionality is desired. A significant goal (but not a requirement) of the MIDS program is for terminals to be interoperable at the card or SEM-E level—that is, to have the ability to take a card from the MIDS terminal on a European aircraft, place that card into a MIDS terminal on a U.S. aircraft, and have that terminal function correctly. If this interoperability goal is realized, it could increase the logistics flexibility of NATO allies' aircraft equipped with MIDS terminals.

The MIDS LVT software architecture is divided into two major parts, as shown in the right-hand diagram of Figure C.1. The common software core supports basic functions such as message processing, signal processing, and Link 16 waveform generation. This part is employed in the basic functioning of all MIDS terminals on all MIDS platforms.

The other major part is the I/O software module, which contains specific I/O interfaces for each LVT platform. For example, the U.S. reference platform for the LVT is the U.S. Navy F/A-18. The LVT I/O software module contains all the necessary software interfaces for the LVT to reside on the F/A-18 avionics bus and exchange information with other relevant systems on this bus. This module has grown in size and complexity because it contains the I/O interfaces needed for all LVT platforms. By design and to ensure compatibility, each LVT terminal is loaded with the same I/O software module, although only a portion of the module is used on any specific LVT platform. Thus, as the number of LVT platforms increases, so does the size of the I/O module. The LVT I/O module has grown to be on the order of 340,000 lines of code. Similarly, the LVT performance specification has grown to be over 800 pages in length, and the interface control document is now over 1500 pages long. Thus, a key challenge for the MIDS program from the beginning has been the harmonization of terminal requirements for the LVT platforms of the participating member nations. In the production phase, the MIDS LVT terminal will be manufactured by a U.S. and European industrial team.[6]

---

[6]For a more detailed discussion of MIDS LVT requirements issues that were negotiated between the international member nations, see Gonzales et al. (2000).

## MIDS LVT PLATFORMS

Originally, the LVT was to be installed only on the U.S. Navy F/A-18, on U.S. Navy ships, and on the ships, aircraft, and ground C2 centers of the European MIDS participants (France, Germany, Italy, and Spain). The U.S. Air Force and Army were not program participants. In 1994 the U.S. Army decided to procure a version of the MIDS terminal called LVT(2). The LVT(2) terminals have different physical characteristics but are operationally interoperable with other Link 16 terminals, including LVT(1), which is the designation for the earlier LVT terminal. In 1998 the Air Force decided to acquire MIDS LVT(1) terminals for the F-16 and ABL. At the end of 1999, the LVT(1) and the LVT(2) were to be integrated into a large number of platforms (see Table 9.1), including the U.S. F/A-18, F-16, Spanish EF-18, Italian Tornado, French Rafale, and EF-2000 (Typhoon) aircraft.

In light of lessons learned from recent joint experiments and combat operations, the number of U.S. aircraft equipped with MIDS is likely to grow in the future. According to the Air Force road map, several additional aircraft are scheduled to receive Link 16 terminals. These aircraft are listed in the last row of Table 9.1. It should be noted that in late 1999, efforts were under way in Congress to procure Link 16 (MIDS LVT) terminals for Air Force B-1, B-2, and B-52 aircraft and Navy EP-3 and S-3 aircraft as well as to acquire additional LVT(2) terminals for the Army.

### FDL Characteristics

In 1995, at the direction of the OSD, the Air Force joined the MIDS program with a new variant of the MIDS LVT called the MIDS Fighter Data Link (FDL).[7] The FDL terminal is designed specifically for the U.S. Air Force F-15. FDL is 80 percent common in hardware and software with the LVT. It shares the same modular architecture as the LVT, although the FDL chassis is a slightly modified version of the LVT chassis. However, new variants of existing LVT components were necessary in some cases because of unique avionics standards or performance requirements related to the F-15.

---

[7]The origin of the FDL terminal procurement is discussed later in this case study.

The FDL terminal differs in design from the LVT. The FDL has no voice or Tactical Air Navigation System (TACAN) capability and has a maximum transmit power level of 50 watts, in comparison to the 200-watt maximum transmit power level for the LVT. Thus, the FDL has a smaller antijam link margin than the LVT in the transmit mode and a shorter maximum range—200 nmi as compared to the 300 nmi of the LVT. Both the LVT and FDL have the same physical dimensions of 0.6 cubic ft.

## HISTORY, SCHEDULE, AND MANAGEMENT STRUCTURE

The long and turbulent histories of the JTIDS and MIDS programs are illustrated in Figure C.2. In the late 1960s, the Air Force and Navy began JTIDS technology developments. The Air Force JTIDS system was based on a time division multiple access (TDMA) architecture, while the Navy's competing JTIDS system was based on a distributed time division multiple access (DTDMA) architecture. Technical problems hampered both programs early on, but the problems encountered by the Navy were more severe. A working prototype DTDMA terminal was never demonstrated.

In 1974 Dr. William Perry, then director of DoD Defense Research and Engineering (DDR&E), directed that the service JTIDS programs be combined into a single joint program. The JTIDS Joint Program Office was created in 1976, and the Air Force was given the lead for this effort. The first operational application of JTIDS was the U.S. Air

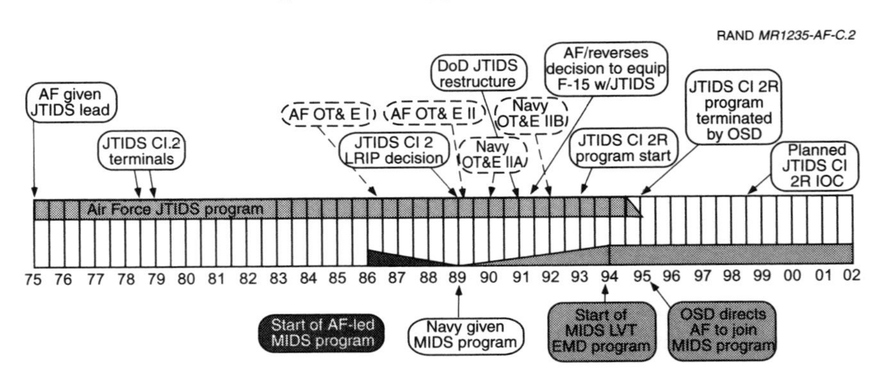


**Figure C.2—Turbulent History of the JTIDS and MIDS Programs**

Force JTIDS Class 1 terminal for U.S. AWACS aircraft. These terminals were large and took up several cabinets' worth of space on AWACS. They could not fit onto smaller fighter aircraft. In the late 1970s the Air Force started efforts to produce a JTIDS Class 2 terminal that could fit within the small confines of fighter aircraft. This effort encountered significant technical challenges, and progress was initially slow.

In the 1980s and early 1990s, a series of operational test and evaluations (OT&Es) were conducted by the Navy and Air Force to evaluate the performance and reliability of JTIDS Class 2 terminals. A number of significant problems were encountered in these operational tests, including high terminal failure rates, short lifetimes of key components, and software reliability problems. In short, the terminals were not reliable. Although the Air Force in 1989 went ahead with a decision for low-rate initial production (LRIP) of the JTIDS Class 2 terminal, DoD later restructured the JTIDS program because of these problems. The Air Force also grew increasingly concerned about the cost and reliability of JTIDS Class 2 terminals, even though there was strong support within the Air Force for putting a data link capability on fighter aircraft. In the face of mounting costs, reliability concerns, and worsening budget pressures, Air Force in 1991 the reversed its decision to equip the F-15 with JTIDS.

After the Gulf War, however, the importance of data communications for situational awareness and for the rapid transfer of targeting and threat information became apparent. In the early 1990s, the Air Force conducted a series of successful operational tests of candidate JTIDS Class 2 terminals with an F-15C squadron at Mountain Home AFB. In 1993 the Air Force started the JTIDS Class 2R program, with the Air Combat Command publishing the operational requirements document in 1994.

## NATO, Link 16, and MIDS

In parallel with United States national efforts to develop the JTIDS Class 2 terminal, the United States and NATO engaged in diplomatic and programmatic efforts to promote interoperability between NATO allies. In 1976 Dr. William Perry offered JTIDS to NATO. NATO interest in JTIDS waxed and waned over the next decade. In 1987 NATO signed a Military Operational Requirement (MOR) document stating the need for jam-resistant tactical communications. In that same year the North Atlantic Council directed the Conference of National Armaments Directors to complete the NATO STANAG on MIDS (STANAG 4175, *Characteristics of MIDS*). Although the NATO MOR remains in effect, NATO did not start an acquisition program for a Link 16–capable terminal.

Instead, a number of NATO member nations initiated an R&D program in 1987 to develop a MIDS terminal in conformance with STANAG 4175. Initially, the Air Force led the U.S. portion of the MIDS program. In 1989, however, the Air Force became increasingly concerned about the reliability and cost of JTIDS Class 2 terminals. Shortly after, the Air Force withdrew the F-16 as the U.S. reference platform for the planned MIDS terminal, and the U.S. Navy quickly responded by offering the F/A-18 as the new U.S. MIDS reference platform. The Navy proposal was accepted by OSD, and the Navy assumed leadership of the MIDS program in early 1990. It should be noted that under Navy leadership, MIDS has never been a joint-service program. Since 1990, the MIDS program has been led by the U.S. Navy, and the IPO is at SPAWAR PMW-101.

A MIDS PMOU was signed by the participating member nations in 1991. This document places a restriction on member nations and forbids them to develop "competing systems" to the MIDS terminal. PMOU Supplement 1 (S1), also signed in 1991, authorized pre-EMD negotiations among participants and initial risk reduction activities. In early 1994, after the program had passed Defense Acquisition Board (DAB) review, the U.S. Navy was authorized to sign PMOU Supplement 2 (S2) and to award an EMD contract to MIDSCO, the consortium of international companies authorized to bid on the program.[8] PMOU S2 defines the cost shares and management structure for the EMD program and gives the MIDS IPO the authority to contract directly with MIDSCO. It also establishes EMD exit criteria (i.e., criteria for successful completion of this phase of the program).

---

[8]MIDSCO is made up of GEC-Marconi Hazeltine (U.S.), Thomson-CSF (France), MID (Italy), Siemens (Germany), and ENOSA (Spain).

## Fate of the JTIDS Class 2R Terminal

Meanwhile the U.S. Air Force and its industry partners were proceeding smoothly with the development of the JTIDS Class 2R terminal. The Air Force issued a cost target of $100,000 per terminal and planned a commercial off-the-shelf/nondevelopmental item acquisition program. The Class 2R terminal was comparable in size and weight to the LVT, although it did not have all its capabilities (i.e., no voice or TACAN capability and lower power). In 1993, the Air Force determined that it had an urgent operational requirement for a Link 16 capability on its air-superiority fighters, the F-15, by the end of 1998.

Shortly thereafter, however, OSD became aware of the conflict between the MIDS LVT and the Air Force JTIDS Class 2R terminal programs. In 1995, to maintain compliance with the MIDS PMOU, OSD directed the Air Force to terminate the Class 2R program and effectively to join the MIDS program. A compromise was reached that satisfied the restrictions of the MIDS PMOU and that could apparently satisfy the Air Force's urgent need for an F-15 Link 16 capability by the end of 1998. The Air Force was permitted to proceed with the acquisition of a Link 16 terminal for the F-15, the FDL terminal, but the MIDS IPO was given responsibility for the acquisition of FDL (also called LVT((3)). FDL was selected over LVT (now designated LVT((1)) for the F-15 because of its expected earlier availability and lower costs. In addition, the FDL program was authorized to enter directly into the production phase and to bypass the MIDS LVT EMD program—at least in terms of acquisition milestone decisions—in order to meet the urgent operational requirement alluded to above.

Despite the testing and acquisition problems associated with JTIDS Class 2 terminals designed for fighter aircraft, JTIDS is operational on U.S. Air Force, U.K., French, and NATO AWACS; on JSTARS, Rivet Joint, ABCCC, the E-2C, the F-14, and the Army Patriot; and on Navy ships.

## LVT Engineering and Manufacturing Development

We now turn to the history of the LVT EMD program and to the prognosis for a smooth transition to the LVT production phase. EMD started in 1994 with a six-month restructuring and development study in order to reduce program cost and schedule by implementing a streamlined acquisition approach, as indicated in Figure C.3. At that time, the MIDS member nations agreed to an open architecture and the use of commercial parts. A one-year reduction in the original EMD program schedule was planned in this streamlining effort. In the third quarter of 1995, the Army LVT terminal, LVT(2), was added. The Air Force FDL LVT(3) contract was awarded in the third quarter of 1996.

Beginning in 1999, the MIDS IPO announced significant delays in the EMD program. Terminal LRIP was delayed one year to the first quarter of CY 2000. The Milestone III decision was delayed two years to the beginning of CY 2002. EMD, originally scheduled to end in CY 1999, was extended six months in order to meet the program exit criteria. In addition, production readiness activities were extended by six months.

According to the MIDS IPO, there were two major reasons for the delays in the EMD program. The first was the lack of a sufficient number of EMD terminals for terminal-platform integration activities. The second reason for the delay was the slow pace and incremental delivery of the TDP.

As of April 1999, 33 EMD terminals were to be produced and made available to the member nations, and the first production terminals

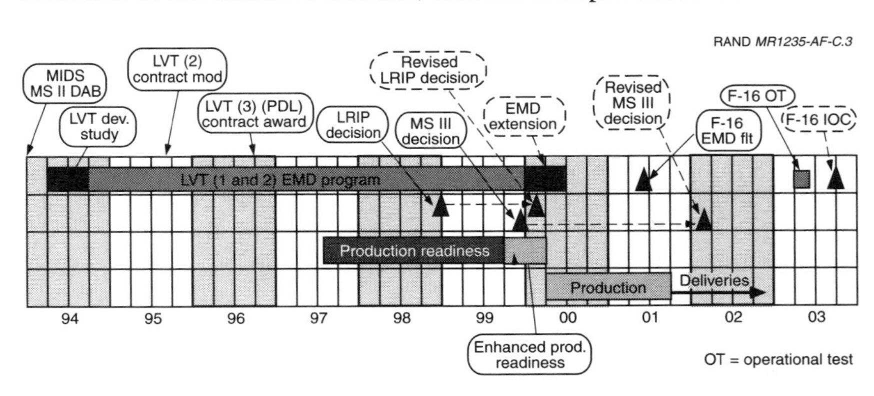


Figure C.3—MIDS LVT Program History and Prognosis

were to be available in CY 1999. However, only 19 were available and the terminal production phase was delayed until CY 2000. There appear to be two reasons for the EMD terminal shortage. The first is a shortage of key parts from foreign suppliers. The second is requirements growth, which has delayed final terminal design. A complicating factor was the late addition of the U.S. Air Force F-16 to the LVT EMD program. This increased the EMD terminal requirement substantially beyond the initial 33 that were contracted for.

A complete TDP is a critical deliverable of the EMD program—it is essential for ensuring competition and contractor readiness for the production phase. It will also be critical for ensuring competition for the U.S. production contract. The TDP will be owned by the MIDS member nations and not by MIDSCO, so the entire TDP or portions of it could be made available to U.S. contractors that are not members of MIDSCO. It will not provide a build-to-print blueprint of the EMD terminal; however, it should provide sufficient technical detail to produce many of the system components. The status of the TDP has caused concern in some circles regarding LVT contractor production readiness and the MIDS IPO production plan.

From a U.S. Air Force perspective, the immediate effect of the delays described above has been to move the IOC for the first Link 16–capable F-16 squadron to the third quarter of CY 2003. This represents a delay of almost three years in fielding an operational data link capability for the F-16. In addition, although its design has been approved by the MIDS IPO and MIDS Steering Committee, the F-16 LVT interface has yet to be implemented. There is concern within the Air Force that the necessary software for this interface will not be completed before the scheduled end of the EMD program. This could further delay the F-16 MIDS LVT IOC and could thus cause further delays in various operational capabilities and significant reprogramming actions by the F-16 SPO.

## FDL Production Program

Initially, the design of the FDL terminal for the F-15 was significantly different from that of the LVT. This is because the programs were developed in response to different requirements. In contrast to LVT, FDL has a lower-power transmitter and lacks TACAN and voice capability. However, there has been substantial sharing of technology and system design information between the two programs, and over the course of time the designs of the two terminals have converged. From the program management standpoint, the relationship between the two programs has been informal even though many contractors are developing and producing components for both terminal systems.

When the FDL program began, the scheduled date for the IOC of the first F-15 squadron was at the end of 1998 (which coincides with the original requirement for the JTIDS Class 2R terminal IOC), as indicated in Figure C.4. However, according to MIDS IPO projections, the FDL IOC will occur approximately 22 months later than the planned IOC for the JTIDS Class 2R terminal.[9] If additional parts delays are encountered in the LVT program, there could be further delays in the FDL program, which could delay IOC and force significant reprogramming actions by the F-15 SPO.

## MIDS IPO Management Structure

During the EMD phase of the program, the MIDS IPO is headed by a U.S. Navy O6-level program manager and a French O6-level officer

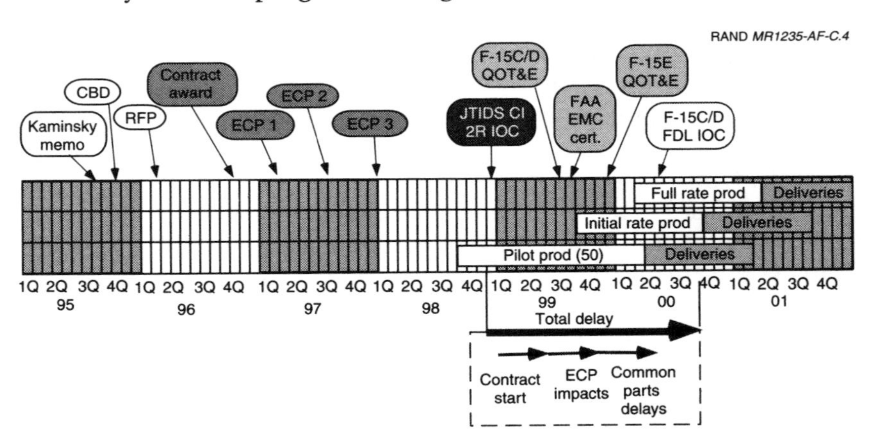


Figure C.4—MIDS FDL Program History and Prognosis

[9]For more detailed information on the causes of the delay, see Gonzales et al. (2000).

deputy program manager. Program oversight is provided by an international board of directors called the MIDS Steering Committee, which consists of one representative from each country. The U.S. representative is the OSD program sponsor, currently the Director of Command, Control, and Communications within OASD(C3I). The Steering Committee is reportedly given substantial management authority for the EMD program as stipulated in PMOU S2.

The MIDS EMD program also has a U.S.-only management committee, the PEC. The current U.S. Air Force representative to the PEC is the ESC Vice Commander. The PEC is chaired by the Navy Program Executive Officer (PEO). To date, the role of the PEC has been largely limited to financial issues such as program cost sharing between the individual services. As noted above, MIDS is not a joint-service program—it is a Navy-led international program. Therefore, according to the current MIDS program manager, detailed joint issues regarding costs and schedules are resolved within the MIDS IPO.

At the time this research was conducted, the senior Air Force officer within the MIDS IPO was an O6-level officer. However, this individual did not have any written or agreed-upon responsibilities within the IPO. He was essentially an Air Force representative-at-large within the IPO. *Therefore, it is not clear that the senior Air Force officer in the IPO has had timely access to all program information necessary to manage the program efficiently from an Air Force perspective or to coordinate with other acquisition organizations in the Air Force, such as the fighter SPOs.*

The management structure for the EMD program was established several years ago, when PMOU S2 was negotiated and signed by the member nations. However, the management structure and cost shares among the international partners for the production phase are now being negotiated and will be established in PMOU S3. Details of the draft PMOU S3 were not made available to the authors except for a few major features of the draft agreement. One of these stipulates that the IPO will remain the central management structure for the procurement of both European and U.S. MIDS terminals. European MIDS terminals will be acquired under a separate contract with a single European industry consortium. However, the U.S. contract for MIDS terminals destined for U.S. platforms will be competed. Many of the details of how these two contracts will be managed are likely to be laid out in PMOU S3.

A second agreement, the JMOA, establishes the contributions and roles of the individual U.S. military services in the MIDS program. A draft JMOA for the production phase of the MIDS program was also under negotiation at the time this research was completed in late 1999. Because the EMD program is essentially a Navy program, the roles of the Air Force and Army in the program are rather limited. An issue we recommend that the Air Force address in the ongoing negotiations over the production-phase JMOA is whether it should have a larger and more substantial role in the production phase to ensure that the MIDS IPO establishes and maintains a viable production plan that will deliver MIDS terminals on schedule and within budget, and that effectively supports Air Force fighter and bomber SPO modernization needs.

## MIDS AND CLASS 2R COSTS

In this section we consider the cost implications of achieving interoperability among NATO allies with the MIDS program. First, we briefly review the cost structure of the current EMD program. Next, we consider how the costs of the FDL terminal would compare with those of the JTIDS Class 2R terminal if the latter program had not been canceled. This enables us to assess the cost penalty, if any, for achieving interoperability in a cooperative development program such as MIDS.

RDT&E funding shares, defined in the PMOU S2, were determined in rough accordance with their expected share of the total buy and the value of the EMD contracts let to contractors from participating MIDS program member nations (the work share allocated to each country). The largest share has been funded by the United States at approximately $265 million, with France and Italy not far behind.

As noted earlier, in February 1995 the U.S. Air Force announced the need for a reduced-function JTIDS Class 2 terminal (the Class 2R) for the F-15 to meet a critical-need date of December 1998. It subsequently received a bid to provide Class 2R terminals at a unit cost of $109,000 (FY 1999 dollars). In August of that year, the Under-secretary of Defense for Acquisition & Technology directed the Air Force to join the MIDS LVT program and to cancel the Class 2R.

The cost implications of the decision to cancel the Class 2R and join the MIDS program are uncertain, but the range of possible cost impacts for this decision is shown in Figure C.5. The current not-to-exceed (NTE) cost for the pilot buy of 50 MIDS FDL terminals is about $183,000. These are the only units under contract at this writing. The objective cost is approximately $160,000. Note that both the Class 2R target and FDL objective costs are approximate. The Class 2R terminal was never under contract, and the FDL costs for the remaining lot buys remain to be negotiated.

Had the Air Force been able to procure the Class 2R at the target price, the additional cost per terminal for FDL would have been $50,000 to $75,000, or between $13 and $20 million for the current 257-terminal buy, enough to equip about 70 percent of the F-15C/D

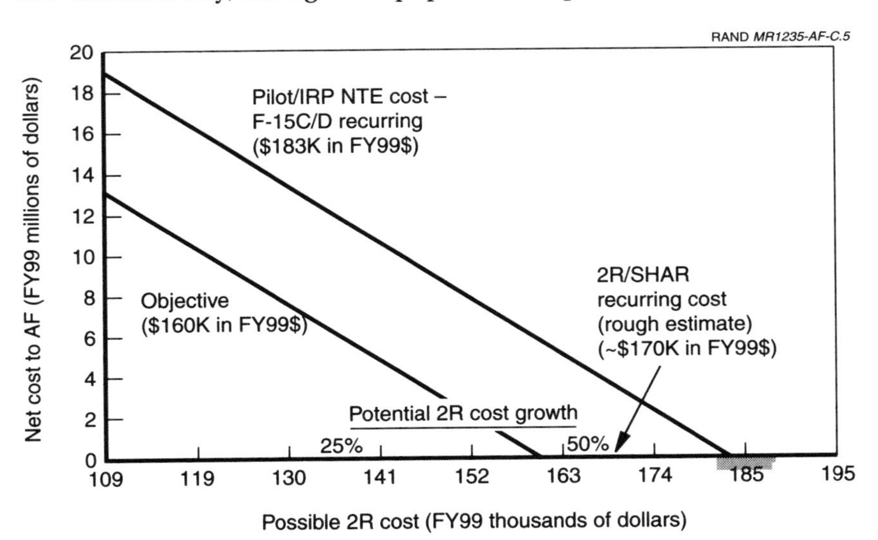


**Figure C.5—Cost Implications of the Decision to Cancel the JTIDS Class 2R Program and Buy FDL**

fleet. This $13–$20 million figure represents an upper bound on the net cost impact of the decision to cancel the Class 2R and procure MIDS for the F-15.

However, it is not clear that the Class 2R program would have come in at its target cost. The Rockwell-Collins SHAR terminal is a derivative of the Class 2R terminal, and its specifications are similar to those of the original Class 2R. Thus, it provides an indication of what the Class 2R might have cost had it continued in development and gone into production. The SHAR cost is consistent with FDL costs, falling midway between the NTE pilot cost and the objective cost. Had the Class 2R come in at the SHAR cost, the net cost of the decision to terminate the Class 2R and buy FDL would have been negligible. Had the Class 2R cost increased by 25 to 50 percent, the net cost impact would have been on the order of a few million to several million dollars.

Even in the worst case, the net cost of the decision to cancel the Class 2R and procure FDL is small relative to Air Force spending on modifications for the F-15, which runs into the hundreds of millions of dollars each year. Had the Class 2R not achieved its cost objective, the net cost impact would have been even less.

Adding in RDT&E spending provides a more complete assessment of the possible cost impact of the decision to cancel the Class 2R terminal and procure FDL (see Figure C.6). As noted earlier, the Air Force may have incurred as much as $20 million in additional costs from that decision. Cost growth in the Class 2R program could have reduced this total considerably.

However, procuring a MIDS terminal variant (FDL) for the F-15 allowed the Air Force to leverage the $650 million RDT&E investment made by the U.S. Navy and the allied partners involved in the program. This investment level dwarfs the relatively modest RDT&E effort associated with FDL. The high degree of commonality between the FDL and LVT terminals and software suggests that the FDL program has benefited from LVT RDT&E. Discussions with the program office confirm this finding.

Had the Air Force continued with the Class 2R program, it might have been forced to fund additional development efforts, thus offsetting some of the additional costs associated with procuring

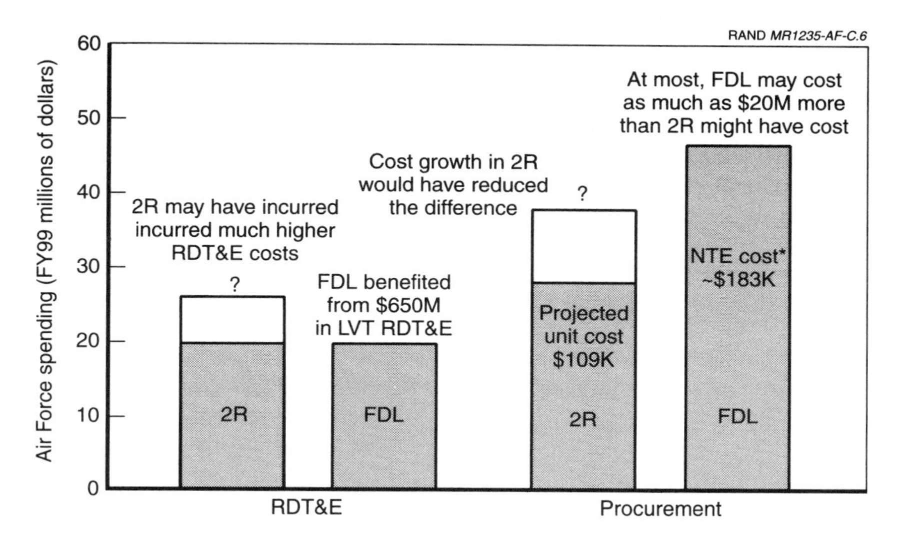


**Figure C.6—Additional Costs of FDL May Be Limited**

FDL. The Class 2R terminal did not continue in development, so however, so it is impossible to know what additional RDT&E costs might have been incurred. Still, any additional costs would have further reduced the net cost of the decision.

In some ways, the MIDS terminal program has been a dramatic success. Six nations have participated in the development, and all plan to procure the system. Each of the participants has a substantial incentive to continue with the program, as their domestic industries are rewarded as a function of the country's participation. The integration of MIDS on allied fighters should substantially increase situational awareness and subsequently increase force effectiveness. U.S. exercises have repeatedly shown that JTIDS-equipped aircraft perform much more effectively than equivalent aircraft that lack the system.

However, the success of the program has not come without cost. Had the United States and other nations chosen to procure an off-the-shelf JTIDS terminal such as SHAR, they might have avoided paying hundreds of millions in MIDS RDT&E costs. They might have saved on procurement as well, as SHAR's costs are substantially lower than those of MIDS.

The discussion of the MIDS and SHAR programs is complicated by the fact that they are not independent. SHAR is built by Rockwell-Collins, which is a major partner in Digital Link Solutions, a prime contractor on the MIDS program. SHAR probably benefited from the substantial RDT&E investment made in the MIDS program. In the absence of MIDS, SHAR's development costs might have been much greater.

While it is possible that the United States could have procured JTIDS terminals at a lower cost than that of MIDS, such an arrangement would probably have involved only U.S. contractors. The NATO allies may not have been willing to simply procure a U.S. system; they might have pursued an independent program instead and subsequently ended up with a system that might have been less compatible with the U.S. JTIDS systems. Absent U.S. leadership, they might not have pursued a JTIDS capability at all and would subsequently not have achieved the improvements in interoperability that they are likely to enjoy over the next few years.

Thus, the success of the program must be viewed in the context of the broader policy objectives. MIDS may be more expensive than some of the alternatives, but it has created incentives for allied participation and significantly enhanced the likelihood that aircraft participating in future coalition operations will have greater situational awareness—and subsequently greater effectiveness—than they do today.

Appendix D

# NOTIONAL FIGHTER DEPLOYMENT

The withdrawal of many U.S. Air Force units from overseas bases and their redistribution to CONUS locations have made force deployments to theaters of operations more stressful as well as time-consuming. When fighters deploy to a theater, over-the-water flight from CONUS to theater locations and en route refuelings are required. In such deployments, the availability of emergency divert fields and the use of U.S. tankers are essential. Further, overflight rights and basing support granted by allies are important to minimize stress and operational risk.

This appendix documents an analysis of the impact of the preceding factors on deploying additional fighter forces to SWA to augment in-place fighter forces as part of a notional halt-the-invasion operation analyzed in Chapter 11.

## METHODOLOGY AND ASSUMPTIONS

For this analysis, we assume that a notional force of 294 fighters and 150 supporting airlifters are deployed from CONUS (four squadron equivalents from the Mountain Home AFB area and the rest from Shaw AFB in South Carolina and Seymour Johnson AFB in North Carolina) to the Middle East with a nominal destination of the Doha, Qatar, area. The fighters are deployed in squadron-size packages of 18 or 24 aircraft. From nine to twelve airlifters, depending on squadron size and composition, fly from each squadron's home base and precede each squadron by two hours at en route stops and at the destination.

In our analysis, we considered two alternative routing schemes: one representing a notional deployment with NATO partners' support and the other without the support of NATO partners or any other nation.

1. In the supported case,

   - The fighter squadrons notionally positioned in the Shaw and Seymour Johnson area are assumed to fly a great-circle route along the eastern coast of the United States and Canada, over the North Atlantic Ocean and England, and into air bases in Germany (nominally, Bitburg). Following refueling and a crew rest period of 17 hours, they fly over eastern France into the Mediterranean (to avoid overflight of Switzerland and Austria), over Sicily into the Eastern Mediterranean Sea, and over Egypt, the Red Sea, and Saudi Arabia to air bases in the Doha area. This route provides several divert fields in case of emergencies during the open-ocean leg from the eastern United States to Germany. A similar route was used routinely during Operation Desert Shield and other deployments of CONUS fighters to SWA. The distance of this route is about 6845 nmi.
   - The fighter squadrons nominally based in the Mountain Home area fly to the Shaw area, remain overnight, and fly the same route as above. Their airlifters fly a great-circle route over the Northern United States and Canada to Germany and then, on the second leg, along the same route discussed above. This route is about 7660 nmi.
   - The A-10 squadrons deploy from the Shaw area to Lajes, remain overnight, fly to a Sicilian airfield, remain overnight, and finally fly to the Doha area. Without an autopilot and at over a quarter slower airspeed than the fighters, three legs versus two are postulated as reasonable to minimize aircrew stress. Having to overlap the fighter flow of two squadrons per day to arrive immediately following the last fighter squadron increases the tanker requirements for the mission.

2. In the nonsupported case,

   - The fighter squadrons fly from the Shaw and Seymour Johnson area via a great-circle route over the Atlantic Ocean through the Strait of Gibraltar, over the Mediterranean Sea, and over Egypt, the Red Sea, and Saudi Arabia to air bases in the Doha area. This route poses more operational risk because of aircrew fatigue and the lesser availability of divert fields while flying over the Atlantic Ocean. However, a similar route has been used on occasion by individual units (squadrons).[1] The distance of this route is about 6700 nmi.
   - The notional Mountain Home fighter forces fly to the U.S. East coast, remain overnight, and then fly the route of the Shaw forces above. The airlifters fly a great-circle route across the United States, over the Atlantic Ocean, through the Straits of Gibraltar, and then along the same route as above. The distance of this route is about 7635 nmi.

Another key distinguishing factor between the two cases, besides the availability of divert fields, is that the nonsupported case requires fighter aircrews to fly continuously for about 16.7 hours. This is doable but stressing for the aircrews, whereas the routes for the supported case are divided into two shorter flight segments (9 and 7.5 hours). In addition, without the stopover in Germany, it is difficult to compensate during the long nonstop flights for any problems that may occur at the destination airfields (e.g., landing clearance, delivery of support, allocation of beddown facilities, or munitions availability) or with diplomatic clearances.

Several other factors are important in determining a reasonable flow of fighters from CONUS to SWA. Among them are (1) the number of tanker assets available to provide the necessary refuelings, (2) the physical support infrastructures (air base capacity, fuel, etc.) available en route and at destination airfields and the willingness of host nations to make them available, and (3) the availability of weapons and specific support required for fighter operations in theater locations. In general, these factors can vary widely depending on

---

[1]A similar route was flown for initial fighter deployments during Desert Shield (Allister, 2000).

location. In particular, the physical infrastructure of the United States, NATO allies, Saudi Arabia, and Qatar are capable of supporting large numbers of fighters, airlifters, tankers, and C3ISR aircraft.

However, in the past, national interests of host nations have limited the availability of their key assets or use of their airspace. For example, in Operations Nickel Grass and El Dorado Canyon, NATO allies refused to permit U.S. forces to transit their airspace or use their facilities while en route to theater; in Operation Desert Shield, Saudi Arabia allowed coalition members access only to selected airfields, which delayed the buildup of air forces; and in Operation Desert Fox, both Saudi Arabia and Turkey, where the United States had the largest concentration of deployed assets, denied U.S. requests to launch strikes from their territory, forcing the United States to develop alternate plans.[2]

Because of the preceding factors and because the focus of our study is to determine the effect of support from the United States' NATO allies, we chose to examine the notional fighter deployments under two conditions. The first case (the supported case) uses the CONUS-through-Germany route with support from NATO allies (with the notional deployment force listed in Table D.1), with a flow of two

**Table D.1**

**Notional U.S. Air Force Fighter Deployment Force (Year 2010)**

| Type | Quantity[a] |
|---|---|
| F-117 | 18 |
| F-22 | 48 |
| F-15C | 18 |
| F-15E | 66 |
| F-16CJ | 18 |
| F-16CG | 84 |
| A-10 | 42 |
| Total | 294 |

[a]There is a mix of 18- and 24-aircraft squadrons.

[2]For a more complete discussion of a range of issues related to en route and in-theater access and basing, see Shlapak et al. (1999).

fighter squadrons per day into theater. The second case, the nonsupported case, uses the CONUS-through-Gibraltar route (i.e., no support from NATO or North African allies), with the flow of forces into the theater constrained to one fighter squadron per day. Given that the flow of fighter squadrons was set for the two cases, the calculations included determining the arrival time of the first squadron and the number of tankers and airlifters needed to support the deployment.

For the supported flights, we postulate that all fighters except the F-117 are deployed in flights of six with a single drag tanker.[3] For nonstop flights in the nonsupported case, F-22s and F-15s are deployed in flights of six, F-16s in flights of four, and F-117s in flights of two, with all flights deploying with a single drag tanker. In both cases, the flights in an 18/24-fighter package are clustered within an hour block for refueling management. Following arrival in theater, the drag tankers are used to support theater operations. Airlifters are assumed to fly in flights that permit the servicing of multiple airlifters by tankers in one-hour periods. The number of airlifters varies between nine and twelve for each package: two C-5A/B/Cs, one C-17 (except two for 24 A-10s), and four to eight C-141s; this is a mix that applies the same general ratio of total primary aircraft authorized (PAA) by aircraft type to each package. They are assumed to carry from 320 to 430 short tons (st) of cargo—a postulated minimal initial support slice requirement for each 18/24-fighter package.[4] Cargo capacities used for the long flights are C-5A/B (67 st), C-17 (47 st), and C-141 (26 st).[5]

For the supported deployment, we used KC-10s from McGuire AFB, New Jersey, and KC-135s stationed or positioned at Seymour Johnson, Pittsburgh, McGuire, Niagara Falls, Pease, and Bangor in the eastern United States; Fairchild, Salt Lake, Lincoln, Grand Forks,

---

[3]The F-117 is currently limited to a ratio of only two F-117s per tanker (Office of the Secretary of the Air Force, 1998, p. 19). This is true for both the supported and nonsupported cases.

[4]Based on sampling of actual deployments: 24 F-16s (3- and H-series equipment) and 22 A-10s to developed airfields—333 st and 420 st, respectively; 18 A-10s to a bare base—488 st (John Surovy, Operations Noble Anvil TPFDD, 1999) and 24 F-15Es to a developed airfield—421 st (4th Fighter Wing).

[5]These values are from the Air Force Scientific Advisory Board (1997), p. 60.

O'Hare, and Selfridge in the Midwest; and Mildenhall, Fairford, Rhein Main, and Ramstein in Europe. Use of KC-135RTs (the refuelable KC-135s) is not required. For the nonsupported case, tankers are assumed to be based at the CONUS airfields listed above and in SWA. In addition, the total fleet of KC-135RTs and the rapid turnaround and return of some drag tankers from Doha would be required to meet a one-squadron-a-day schedule.

In both the supported and nonsupported cases, Pease, New Hampshire, was used extensively and almost to excess in the nonsupported case because of its maximum on the ground (MOG), its mission as an Air National Guard base, and the fact that it is the eastern most military airfield within CONUS. We realize that significant ground support from active forces and additional airlift would be necessary to achieve a high-tempo operation at Pease for augmentation tankers.

Parameters for aircraft performance were drawn from *Air Mobility Planning Factors*.[6] Comparisons with some of the technical orders for the aircraft involved showed the fuel burn parameters to be conservative, so these parameters were used for hourly burn throughout without computing climbs and descents.[7] To simplify our calculation of the refueling burden, we calculated hourly fuel consumption for fighters, airlifters, and tankers that were refueled. Fighters were assumed to be refueled hourly by nondrag tankers or by accompanying drag tankers along the route (rather than at preplanned tanker orbits and dedicated tanker tracks). Airlifters and drag tankers do not require frequent (hourly) refueling and were thus refueled during hours when the radii of tanker flights from takeoff airfields to hourly points on the en route tracks of the packages were minimal and offload was most efficient. Fuel required for fighters, airlifters, and tankers included fuel to the destination plus 40 minutes for divert to an alternate field and about eight minutes at the alternate field.[8]

---

[6]See Office of the Secretary of the Air Force (1998).

[7]Technical Orders 1F-15E-1 and 1F16C/D-1.

[8]There are sufficient alternate airfields within 40 minutes (or 250 nmi) of Bitburg and Doha.

Movement of refueling operations from the assumed hourly points would require additional tanker offload sorties (sorties during which fuel offload occurs). Tanker and airlift aircrews were augmented and assumed available for 24 hours without formal aircrew rest, and minimum tanker turnaround times were included when needed.

Note that this analysis was not intended to be an accurate representation of the very complex planning for an actual operational deployment as it would be conducted today, but rather to highlight the potential *relative difference* between two levels of allied support.

## ANALYTICAL RESULTS

Arrival times in the Qatar area for each 18/24 package are shown in Table D.2. Fighter deployment from CONUS to theater in the supported case takes about nine days (eight days without the A-10s), with the first squadron arriving in 62 hours, and requires a total of 243 tanker offload sorties using 65 tankers (including spares).[9] The use of KC-135RTs is not required.

**Table D.2**

**U.S. Fighter Arrival Times in the Qatar Area**

| Type | Number | Supported Case Arrival Time (hr) | Nonsupported Case Arrival Time (hr) |
|---|---|---|---|
| F-22 | 24 | 62 | 62 |
| F-117 | 18 | 70 | 86 |
| F-15E | 24 | 86 | 110 |
| F-16CJ | 18 | 94 | 134 |
| F-22 | 24 | 110 | 158 |
| F-15E | 18 | 118 | 182 |
| F-16CG | 18 | 134 | 206 |
| F-16CG | 24 | 142 | 230 |
| F-15C | 18 | 158 | 254 |
| F-15E | 24 | 166 | 278 |
| F-16CG | 18 | 182 | 302 |
| F-16CG | 24 | 190 | 326 |
| A-10 | 24 | 206 | — |
| A-10 | 18 | 214 | — |

[9]The two squadrons of A-10s require almost 60 hours en route, so if they are to arrive on the day following the fighters, they cost additional tanker availability.

In the nonsupported case, it takes almost 14 days to complete deployment and requires a total of 504 tanker offload sorties using 210 tankers (including spares). Because of the direct route, fighters could arrive in 42 hours and all subsequent squadrons 20 hours sooner than shown in Table D.2; however, for this comparative analysis, first arrival time was held the same as for the supported case (62 hours). The total fleet of KC-135RTs would be required to meet a one-squadron-a-day schedule. A-10s are unable to participate because of excessive aircrew fatigue from flying over 24 hours without en route stops.

# BIBLIOGRAPHY


Adkin, Keith (ed.) (1998), *Jane's Electro-Optical Systems 1998-1999,* Jane's Information Group Ltd., Alexandria, Va.

*Aerospace Daily* (1996), "GEC Marconi, RI Team Wins F-15 Datalink Competition," Vol. 180, No. 2, Aviation Week Group, Washington, D.C., October 2, p. 12.

Air Combat Command (1994), *Operational Requirements Document (ORD), CSF 315-92-I/II/III-A, for the Class 2R Data Link Radio, ACAT Level III,* ACC/DR, Langley AFB, Va., December 1.

Air Combat Command (1996), "AWACS Requirements Roadmap," ACC/DRC, Langley AFB, Va., May 16.

Air Combat Command (1997), "Surveillance and Reconnaissance Mission Area Plan," draft, ACC/DRR, Langley AFB, Va., September 17. Government publication; not releasable to the general public.

Air Force Scientific Advisory Board (1997), *Report on United States Air Force Expeditionary Forces, Volume 1: Summary,* SAB-TR-97-01, Washington, D.C., November.

Air Group IV (1999a), "Air Group IV: Intelligence, Surveillance, and Reconnaissance," website, NATO Air Force Armaments Group, updated September 13, http://www.nato.int/structur/AC/224/ag4/ag4.htm.

Air Group IV (1999b), "The Standing Interoperability and Applications Working Group on Intelligence, Surveillance, and

Reconnaissance (SIAR WG)," website, NATO Air Force Armaments Group, updated September 13, http://www.nato.int/structur/AC/224/ag4/siar.htm.

Air Group IV (1999c), "The Standoff Surveillance and Target Acquisition Systems (SOSTAS) Interoperability Ad Hoc Working Group (SI AHWG)," website, NATO Air Force Armaments Group, updated September 13, http://www.nato.int/structur/AC/224/ag4/si.htm.

Aldous, Derek (1999), "U.S. Radar Offer for NATO AGS," *NATO Interoperability Conference*, H. Silver and Associates, London, March 25–26.

Allister, Paul E. (2000), Assistant Command Historian, discussions via electronic mail regarding fighter force deployments during Operation Desert Shield, Air Combat Command, ACC/HO, Langley AFB, Va., January 12.

Army Science Board (1995), *Technical Information Architecture for Command, Control, Communications and Intelligence*, 1994 Summer Study Final Report, April.

Ashworth, Maj. Jim (1999), discussions regarding MIDS acquisition, SAF/AQI, Rosslyn, Va., June 21.

AWACS Interoperability Review Group (1996), "Terms of Reference for the Interoperability Review Group (IORG)."

Barensky, Stefan (1999), "EU/ESA Push For Independent GPS-Like System," *International Space Industry Report*, Vol. 3, No. 5, June, p. 15.

Boone, Lt. Col. Douglas (1999), "Common GMTI Format Working Group," briefing charts presented to working group meeting held at Rome, N.Y., ASC/RAJI, Wright-Patterson AFB, Ohio, November 3.

Borland, Frank (1999), "NATO's Defence Capabilities Initiative: Preparing for Future Challenges," *NATO Review*, Vol. 47, No. 2, Brussels, Belgium, Summer, pp. 26–28 (http://www.nato.int/docu/review/9902-06.htm).

Braybrock, Roy (2000), "JSF: All Things to All Men?" *Air International*, Vol. 58, No. 6, Key Publishing Ltd., Stamford, U.K., June, p. 372.

de Briganti, Giovanni (1995), "German Indecision Grounds French Space Plan," *Space News*, Army Times Publishing Co., Springfield, Va., October 2–8, p. 10.

Brownell, Thomas (1999), Chief, Plans & Evaluation Division, discussions regarding the NATO AWACS program, NATO AEW&C Programme Management Agency, Brunssum, The Netherlands, August 18.

Bruce, Robert (1999), Director, Atlantic Armaments Cooperation, discussions regarding the NATO AGS program, Office of the Deputy Under Secretary of Defense for International and Commercial Programs, Washington, D.C., September 29.

Bull, Norman S. (1998), *Multifunctional Information Distribution System–Low Volume Terminal (MIDS-LVT)—A Case Study Providing a History of the MIDS-LVT Program*, Defense Systems Management College, Fort Belvoir, Va., November.

Burnham, Randy (2000), briefing charts on Link 16, Air Force Tactical Data Link Systems Integration Office (ESC/DIVJ), Electronic Systems Center, Hanscom AFB, Mass.

Caragianis, Lt. Col. George (1999), Chief, Surveillance Branch (C2RS), discussions with C2RS staff regarding the AWACS and JSTARS programs, Aerospace C2 & ISR Center, Langley AFB, Va., June 8.

Chairman of the Joint Chiefs of Staff (1996), *Joint Vision 2010*, Department of Defense, Washington, D.C.

Christen, Col. Craig, USAF (1999), discussions regarding the MIDS program, MIDS International Program Office, San Diego, Calif., February 26.

CNES (1999), "CNES Strategic Plan: Innovation for the Development of Space Applications," white paper presented at George Washington University, Washington, D.C., July 15.

CNES (2000), "France and Germany Together on Military Earth Observation," *France in Space,* No. 120, Office for Science and Technology, Embassy of France, Washington, D.C., June 16 (http://france-science.org/france-in-space).

Cohen, William (1999), Secretary of Defense, "Europe Must Spend More on Defense," *Washington Post,* December 6, p. 27.

Conference of National Armaments Directors (1997), *NATO Staff Requirements (NSR) for an Alliance Ground Surveillance (AGS) System, Final Draft,* Working Paper AC/259(SURV)WP/17 (3rd revise), September 12.

Corbitt, Thomas (1999), NATO technical expert, discussions regarding the NATO ACCS program, NATO Air Command and Control Management Agency, Brussels, Belgium, March 15.

Dahlburg, John-Thor (1999), "NATO Chief Calls for Defense," *Los Angeles Times,* December 3, p. 4.

Department of the Army (1997), "Army Technical Architecture," Version 4.9.5X, draft, July 11.

Department of Defense (1999a), *Report on Allied Contributions to the Common Defense: A Report to the Congress by the Secretary of Defense,* Washington, D.C., March.

Department of Defense (1999b), "Space Policy," DoD Directive 3100, July 9.

Defense Information Systems Agency (1998), *Defense Information Infrastructure Common Operating Environment (DII/COE) Specification,* Washington, D.C.

Deming, Lt. Col. Nancy, USAF (1999), U.S. Research & Development Coordinator, discussions held at MITRE, Hanscom AFB, on the interoperability of U.S. and NATO C2 systems, NC3A, The Hague, the Netherlands, February 9.

Drozdiak, William (1999), "War Showed U.S.-Allied Inequality," *Washington Post,* June 28, p. 1.

Dundas, Lt. Col. Paul, USAF (1999), U.S. Research & Development Coordinator, discussions regarding the NATO ICC program and other NC3A technology R&D, NC3A, The Hague, The Netherlands, March 16.

Electronic Systems Center (1997), *Air Force Concept of Link 16 Employment,* Version 1, *Link 16 Systems Integration Office (ESC/DIAJ),* Hanscom AFB, MA, October. Government publication; not releasable to the general public.

ERIM International, Inc., and Electronic Systems Command (2000), *Joint Eagle Interoperability Experiment Report—Joint Analysis Center, RAF Molesworth, U.K.,* No. 10015100-42-T, Ann Arbor, Mich., and Hanscom AFB, Mass., January.

Ferster, Warren (1999), "NASA Questions Wisdom of Launching Radarsat-2," *Space News,* Vol. 10, No. 7, Army Times Publishing Co., Springfield, Va., February 22, pp. 1, 26.

Gebhard, Paul R.S. (1994), *The United States and European Security,* International Institute for Strategic Studies, IISS/Brassey's, London.

General Accounting Office (1980), *The Joint Tactical Information Distribution System: How Important Is It?,* report to the Congress by the Comptroller General of the United States, PSAD-80-22, Washington, D.C.

General Accounting Office (1992), *Military Communications: Joint Tactical Information Distribution System Issues,* GAO/NSIAD-93-16, Washington, D.C., November.

General Accounting Office (1999), *NATO: Progress Toward More Mobile and Deployable Forces,* GAO/NSIAD-99-229, Washington, D.C., September 30.

Gompert, David C., Richard L. Kugler, and Martin C. Libicki (1999), *Mind the Gap: Promoting a Transatlantic Revolution in Military Affairs,* National Defense University Press, Washington, D.C.

Gonzales, Daniel, Daniel Norton, and Myron Hura (2000), *Multifunctional Information Distribution System (MIDS) Program Case Study,* DB-292-AF, 2000.

GPS Support Center (2000), "Transition of Selective Availability to Zero," chart, May 2 (http://www.peterson.af.mil/usspace/gps_support/archive/Frontpage/J122SA_off.htm).

Hank, Lt. Col. Heinz G., German Air Force (1998), "Joint Forces Air Component Command (JFAAC): A New Concept within NATO in Support of a Combined Joint Task Force (CJTF)," annotated briefing, Headquarters Allied Air Forces Central Europe (AIRCENT), Ramstein AB, Germany, December 4.

Harper, Gp. Capt. C. N., U.K. Royal Air Force (1998), "Air Power for a New NATO," annotated briefing, Headquarters Allied Air Forces Central Europe (AIRCENT), Ramstein AB, Germany, December 4.

Henderson, Breck W. (1990), "USAF, NATO Invest Heavily in AWACS Electronics Upgrades," *Aviation Week & Space Technology*, Vol. 132, No. 1, McGraw-Hill, Washington, D.C., January 1, pp. 45–50.

Hewish, Mark, and Joris Jansen Lok (1999), "Lords of the Sky," *Jane's International Defence Review*, Vol. 32, Jane's Information Group Ltd., Surrey, U.K., July, pp. 18–25.

Huckins, C. Hampton (1999), senior systems engineer, JSTARS International, discussions regarding the JSTARS International Program and the NATO AGS program, JSTARS Program Office, Hanscom AFB, Mass., July 15.

Huges, David (1995a), "USAF Expects JTIDS Ruling," *Aviation Week & Space Technology*, Vol. 142, No. 23, McGraw-Hill, Washington, D.C., June 5, pp. 26–27.

Huges, David (1995b), "F-15 JTIDS Effort Moved to MIDS Program," *Aviation Week & Space Technology*, Vol. 142, No. 26, McGraw-Hill, Washington, D.C., June 26, p. 24.

Hura, Myron, et al. (1998), *Investment Guidelines for Information Operations: Focus on Intelligence, Surveillance, and Reconnaissance*, RAND. Government publication; not releasable to the general public.

Hura, Myron, et al. (1999), *Information for the Warfighter: Integrating C2 and ISR*, RAND. Government publication; not releasable to the general public.

Hutber, David (1999), discussions with DERA representatives regarding ongoing UK C3ISR initiatives, DERA, Malvern, UK, March 11.

IEEE STD 610.12.

International Institute for Strategic Studies (various years), *The Military Balance*, IISS/Brassey's, London.

Jackson, Paul (ed.) (1999), *Jane's All the World's Aircraft 1999-2000*, Jane's Information Group Ltd., Alexandria, Va.

Jevsevar, Lt. Col. Vic (1999), "NAEW&C Mid-Term Programme," annotated briefing, NATO AEW&C Programme Management Agency, Brunssum, The Netherlands, August 18.

Joint Staff (1999), *DoD Dictionary of Military and Related Terms*, Joint Publication 1-02, Department of Defense, Washington, D.C., March 23, 1994, as amended April 6, 1999.

Joint Strike Fighter Program Office (1998), *Selected Acquisition Report—Program: Joint Strike Fighter*, December.

Juday, Lt. Col. Greg (1999), Chief, Weapon System Requirements and Integration, discussions with program office personnel regarding the U.S. and international AWACS programs, AWACS Program Office, Electronic Systems Command, Hanscom, AFB, Mass., July 16.

Jumper, Gen. John P. (1999), USAFE Commander, "NATO Reorganization," annotated briefing, U.S. Air Forces in Europe, Ramstein AB, Germany.

Kirch, Robert (1999), Project Manager, Joint Services Workstation, discussions regarding JSTARS and the JSWS program, Motorola, Scottsdale, Ariz., August 4.

Larson, Eric V. (1996), *Casualties and Consensus—The Historical Role of Casualties in Domestic Support for U.S. Military Operations*, RAND, MR-726-RC.

Larson, Eric, et al. (1999), "The Interoperability of U.S. and NATO Allied Air Forces—Context, Scope, and Key Issues," RAND. Not releasable to the general public.

Lenk, P. J. (1997), "NATO Alliance Ground Surveillance (AGS) Paris Interoperability Experiment: Report on Preliminary Results," Technical Note 721, NC3A, The Hague, the Netherlands, September.

Lennox, Duncan (ed.) (1999), *Jane's Air-Launched Weapons*, Jane's Information Group Ltd., Alexandria, Va.

Lewis, George (1999), discussions regarding the MIDS FDL terminal, Air Force Tactical Data Link Systems Integration Office, Electronic Systems Command, Hanscom AFB, Mass., May 11.

Lewis, J.A.C. (2000), "France, Germany, and Italy Set to Make Space Pact," *Jane's Defence Weekly*, Jane's Information Group Ltd., Surrey, UK, June 21.

Long, Bruce (1998), Chief Architect, discussions regarding the TBMCS program, Lockheed Martin Program Development Office, Colorado Springs, Colo., December 21.

Lopez, Ramon (1999), "Kosovo Air Campaign Shows Up U.S. and NATO Disparities," *Flight International*, Vol. 156, No. 4700, Reed Business Information, Surrey, UK, October 27–November 2, p. 23.

Lorell, Mark, and Julia Lowell (1995), *Pros and Cons of International Weapons Procurement Collaboration*, RAND, MR-565-OSD.

Martel, Colonel, French Air Force (1999), discussions regarding force compatibility and interoperability with USAF, French Air Staff, Paris, March 17.

McLean, A., and A. Swankie (1998), "Helios 2—Myth or Reality?" *Space Policy*, Vol. 14, No. 2, Elsevier Science Ltd., Oxford, UK, May, pp. 107–114.

Meierhoefer, Maj. Axel (1999), discussions regarding the capabilities of the German Air Force to conduct a variety of air-to-ground missions, German Air Force Flying Training Center, Holloman AFB, N.M., March 3.

MIDS International Program Office (1998), *Selected Acquisition Report—Program: MIDS-LVT*, December.

MITRE (1993), *JTIDS Overview Description*, MTR 8413R2, JTIDS Project Staff, Bedford, Mass., February.

Mock, W. (1999), "NATO AEW&C Programme Overview," annotated briefing, NATO AEW&C Programme Management Agency, Brunssum, the Netherlands, August 18.

Morrocco, John D. (1999), "Raytheon Scores Win in ASTOR Competition," *Aviation Week & Space Technology*, Vol. 150, No. 25, McGraw-Hill, Washington, D.C., June 21, p. 30.

Morrocco, John D. (2000), "NATO Radar Project Aims to Boost Interoperability," *Aviation Week & Space Technology*, Vol. 152, No. 23, McGraw-Hill, Washington, D.C., June 5, p. 87.

Mueller, John E. (1994), *Policy and Opinion in the Gulf War*, University of Chicago Press, Chicago, Ill.

Murphy, Pat (1998), discussions regarding the TBMCS program, Lockheed Martin Program Development Office, Colorado Springs, Colo., December 21.

National Institute of Standards and Technology (1996), *Application Portability Profile (APP): The U.S. Government's Open System Environment Profile*, Version 3.0, NIST Special Publication 500-230, Gaithersburg, Md., February.

NATO (1998), *NATO Handbook*, NATO On-Line Library (http://www.nato.int/docu/handbook/1998/index.htm).

NATO (1999a), "The Alliance's Strategic Concept," NATO Press Release NAC-S(99)65, April 24 (http://www.nato.int/docu/pr/1999/p99-065e.htm).

NATO (1999b), "Defence Capabilities Initiative," NATO Press Release NAC-S(99)69, April 25 (http://www.nato.int/docu/pr/1999/p99s069e.htm).

NATO (1999c), "NATO Airborne Early Warning & Control Programme," NATO Basic Fact Sheet No. 28, NATO On-Line Library, updated September 16 (http://www.nato.int/docu/facts/naewcp.htm).

NATO Consultation, Command, and Control Agency (NC3A) (no date), brochure describing the NC3A, The Hague, The Netherlands.

NATO Consultation, Command, and Control Agency (1999), website for NATO Alliance Ground Surveillance system, The Hague, The Netherlands (http://www.nc3a.nato.int/acdiv/surv/agstb.htm).

*NATO's Sixteen Nations* (1998), Special Edition 1998, "AWACS Takes Off into the 21st Century," Jules Perel's Publishing Co, Uithoorn, the Netherlands.

Nethercott, Gerald (1999), Chairman, NATO Air Group IV Standing Interoperability and Applications Working Group on ISR, discussions regarding the need for a NATO standard for GMTI data, Rome Labs (AFRL/IFEC), Rome, N.Y., December 1.

Ochmanek, David A., Edward Harshberger, David A. Maler, and Glenn A. Kent (1998), *To Find, and Not to Yield—How Advances in Information and Firepower Can Transform Theater Warfare,* RAND, MR-958-AF.

Office of the Secretary of the Air Force (1993), *Gulf War Air Power Survey,* Vol. V, Part I: "Statistical Compendium," Washington, D.C. Series commonly referred to as GWAPS. Government publication; not releasable to the general public.

Office of the Secretary of the Air Force (1998), *Air Mobility Planning Factors,* Air Force Pamphlet 10-1403, Washington, D.C., March 1.

Ohl, Lt. Col. Gottlieb (1999), Senior German Representative, discussions regarding structure and mission of IDCAOCs, IDCAOC, Ramstein AB, Germany, August 10.

Pace, Scott, Brant Sponberg, and Molly Macauley (1999), *Data Policy Issues and Barriers to Using Commercial Resources for Mission to Planet Earth,* DB-247-NASA/OSTP, RAND.

Pearlstein, Steven (1999), "Canada Balks at New U.S. Policy," *Washington Post,* The Washington Post Co., Washington, D.C., August 12, p. E2.

*Periscope* (1996), HORIZON entry in "USNI Military Database" online subscription database, UCG/Periscope, Rockville, Md., December 2 (http://www.periscope.ucg.com/docs/weapons/sensors/airradar/w0000199.html).

*Periscope* (1999), force structure entries of NATO nations' air forces in "USNI Military Database" online subscription database, UCG/Periscope, Rockville, Md., February 1 (http://www.periscope.ucg.com/nations/nato/index.html).

Proctor, Paul (1999), "Aging NATO AWACS Prove Reliable in Balkans Campaign," *Aviation Week & Space Technology*, Vol. 151, No. 9, McGraw-Hill, Washington, D.C., August 30, p. 43.

Rhodes, Carl, and Ted Harshberger (1998), "Fielding and Employing an Effective Halt Force—An Integrated Interdiction Model," briefing, RAND, October. Not releasable to the general public.

Rodero, Joseph (1999), Division Manager, discussions regarding the NATO ICC Program and other NC3A technology R&D, NC3A, The Hague, the Netherlands, March 16.

de Romemont, Col. Emmanuel, French Air Force (1999), discussions regarding the French military and political involvement within NATO, Ministry of Defense, Paris, March 17.

Roose, Pamela (1999), Project Manager, International Ground Segment Systems, discussions regarding the CGS, JSTARS, NATO AGS, ASTOR, and CAESAR programs, Motorola, Scottsdale, Ariz., August 4.

Ross, Joe (1999a), "Coalition Aerial Surveillance and Reconnaissance (CAESAR)," briefing, NC3A, The Hague, the Netherlands, July 8.

Ross, Joe (1999b), manager, Alliance Ground Surveillance, discussions regarding the NATO AGS and CAESAR programs, NC3A, The Hague, the Netherlands, August 10.

Russell, Capt. Tom (1999a), MIDS Program Manager, "MIDS LVT & FDL Status: Brief to MULTI-TADIL Senior Level Review," MIDS International Program Office, San Diego, Calif., April 8.

Russell, Capt. Tom (1999b), MIDS Program Manager, "MIDS LVT & FDL Status: Brief to COMACC," MIDS International Program Office, San Diego, Calif., January 6.

Russell, Capt. Tom (1999c), MIDS Program Manager, discussions regarding the MIDS program, MIDS International Program Office, San Diego, Calif., February 26.

Sauer, Rich (1999), discussion regarding JTIDS 2R and MIDS programs, Aerospace C2 & ISR Center, Langley AFB, Va., February 18.

Schaake, Lt. Col. Al, USAF (1999), discussions regarding NATO strategic and operational issues, SHAPE, Mons, Belgium, March 9.

de Selding, Peter B. (1995a), "About Face: Spain, Italy Join Helios 2 Program," *Space News*, Army Times Publishing Co., Springfield, Va., January 16–22, pp. 1, 21.

de Selding, Peter B. (1995b), "Germans Chafe at Helios Tab," *Space News*, Army Times Publishing Co., Springfield, Va., January 23–29, pp. 3, 21.

de Selding, Peter B. (1995c), "WEU Seeks Role as Full Partner in Helios Project," *Space News*, Army Times Publishing Co., Springfield, Va., November 27-December 3, pp. 4, 21.

de Selding, Peter B. (1999), "Italy, Spain Consider Alternatives to Helios 2," *Space News*, Vol. 10, No. 48, Army Times Publishing Co., Springfield, Va., December 20, p. 6.

de Selding, Peter B. (2000), "German Satellite Plan Could Reignite French Partnership," *Space News*, Vol. 11, No. 25, Army Times Publishing Co., Springfield, Va., June 26, pp. 3, 27.

Shlapak, David A., John Stillion, Olga Oliker, and Tanya Charlick-Paley (1999), "A Global Basing Strategy for the U.S. Air Force," RAND. Not releasable to the general public.

Simkol, Joel (2000), "Link-16 Enhancements and TacLink Weapons Brief for Joint Forces Command (JFCOM)," briefing charts, U.S. Navy, SPAWAR PMW-159, San Diego, Calif., March 30.

Solana, Javier, (1999), NATO Secretary General, statement made at the NATO Summit in Washington, D.C., April 24 (http://www.nato.int/docu/speech/1999/s990424a.html).

SOSTAS Interoperability Ad Hoc Working Group (1999), "Terms of Reference for the NATO NAFAG Air Group IV Standoff Surveillance and Target Acquisition Systems (SOSTAS) Interoperability Working Group," draft, June.

Space Publications (1999), *State of the Space Industry 1999*, Bethesda, Md.

Streitmater, Major Kirk (1999a), "Joint STARS," briefing charts, JSTARS Program Office, Electronic Systems Command, Hanscom AFB, Mass., June 23.

Streitmater, Major Kirk (1999b), Chief, JSTARS Advanced Development, discussions regarding the JSTARS program, JSTARS Program Office, Electronic Systems Command, Hanscom AFB, Mass., July 15.

Strazdis, Luke (1999), discussions regarding the JSTARS and NATO AGS programs, NC3A, The Hague, the Netherlands, August 10 .

Streetly, Martin (ed.) (1999), *Jane's Radar and Electronic Warfare Systems 1999–2000*, Jane's Information Group Ltd., Alexandria, Va.

Surovy, John (1999), personal communication regarding "USAFE Operations Noble Anvil Time-Phased Force Deployment Data," working computer file, U.S. Air Forces in Europe, Logistics Plans Directorate, Ramstein AB, German, October.

Synectics Corp. (1998), *Diminishing Manufacturing Sources (DMS) Studies for Joint Tactical Information Distribution Systems (JTIDS)*, ADA 343494, Rome, New York, April.

Talbott, Strobe (1999), Deputy Secretary of State, "America's Stake in a Strong Europe," remarks at a conference on the future of NATO, Royal Institute of International Affairs, London, October 7.

Taverna, Michael A. (1999), "Italy Commits to Galileo, Radarsat," *Aviation Week & Space Technology*, McGraw-Hill, Washington, D.C., April 5, p. 66.

Teal Group Corp. (1999a), *World Military and Civil Aircraft Briefing*, Fairfax, Va.

Teal Group Corp. (1999b), *World Missiles Briefing*, Fairfax, Va.

Tessmer, Arnold Lee (1988), *Politics of Compromise: NATO and AWACS*, National Defense University Press, Washington, D.C.

Tirpak, John A. (1997), "Deliberate Force," *Air Force Magazine*, Vol. 80, No. 10, Air Force Association, Arlington, Va., October.

United Nations (1945), *Charter of the United Nations*, New York, June 26 (http://www.un.org/Overview/Charter/contents.html).

USAFE/SA (2000), "Air War Over Serbia (AWOS) Fact Sheet," Studies and Analysis Directorate, U.S. Air Forces in Europe, Ramstein AB, Germany, January 31.

U.S. Air Force (1996), *Global Engagement: A Vision for the 21st Century Air Force*, Washington, D.C.

Wagner, Caroline (1998), *International Agreements on Cooperation in Remote Sensing and Earth Observation*, RAND, MR-972-OSTP.

WEU Satellite Centre (1998), brochure, May (also available at http://www.weu.int/satellite/en/).

WEU Technological and Aerospace Committee (1999), *Space Systems for Europe: Observation, Communications, and Navigation Satellites: Reply to the Annual Report of the Council*, May 18.

The White House (1994), "Fact Sheet: Foreign Access to Remote Sensing Space Capabilities," discusses PDD-23, March 10.

The White House (1996), "Fact Sheet: National Space Policy," discusses PDD-49, September 19.

The White House (1998), *A National Security Strategy for a New Century*, Washington, D.C., October.

The White House (2000), "Statement by the President Regarding the United States Decision to Stop Degrading Global Positioning System Accuracy," Office of the Press Secretary, Washington,

D.C., May 1 (http://www.peterson.af.mil/usspace/gps_support/archive/Frontpage/SA.doc).

Whiteway, Wg. Cdr. Hilary, UK Royal Air Force (1999), discussions regarding NATO command and control technology initiatives, SHAPE, Mons, Belgium, March 8.

Wininger, Lt. Col. David, USAF (1999a), Deputy Assistant Chief of Staff Requirements & Modernization, discussions with NAEWFC personnel regarding NATO AWACS requirements, operations, and training, NATO AEW Force Command Headquarters, collocated with SHAPE, Mons, Belgium, August 17.

Wininger, Lt. Col. David, USAF (1999b), "NATO AEW Force," briefing charts, NATO AEW Force Command Headquarters, collocated with SHAPE, Mons, Belgium, August 17.

Wolfe, George (1999), senior systems operations officer, discussions regarding the NATO AWACS program, NATO AEW&C Programme Management Agency, Brunssum, the Netherlands, August 18.